Conceptual structure in childhood and adolescence

'Heat breaks up charcoal and puts sulphur dioxide in'; 'The air pulls faster on heavy masses.' These and other similar statements by school-aged children untutored in physics carry two messages. First, children's pre-instructional conceptions of the physical world are a far cry from the received wisdom of science; second, despite their lack of orthodoxy, children's conceptions carry a definite sense of causal mechanism. This sense of mechanism is the focal concern of this book for it raises issues of central importance to both psychological theory and educational practice.

In particular, some psychologists have claimed that human cognition is organised around causal mechanisms along the lines of a theory. This carries specific implications for teaching. Does the existence in children's thinking of causal mechanisms relating to the physical world support these psychologists? Does this have consequences for the teaching of science?

Christine Howe reviews evidence relating to pre-instructional conceptions in three broad topic areas: heat and temperature; force and motion; floating and sinking. A wide range of published work is discussed, including the author's own research. In addition, a new study covering all three topic areas is reported for the first time. The message is that causal mechanisms can indeed play an organising role, that untutored cognition can in other words be genuinely theoretical. However, this tendency is highly domain-specific, occurring in some topic areas but not in others.

Having drawn these conclusions, Christine Howe discusses their meaning in terms of both cognitive development and educational practice. A model is outlined which synthesises Piagetian action-groundedness with Vygotskyan cultural-symbolism and has a distinctive message for classrooms. *Conceptual Structure in Childhood and Adolescence* will be useful to cognitive and developmental psychologists and to science educators alike.

Christine J. Howe is a Reader in Psychology at the University of Strathclyde. Her previous publications include: *Acquiring Language in a Conversational Context* (1981); *Language Learning: A Special Case for Developmental Psychology?* (1993); *Group and Interactive Learning* (1994) and *Gender and Classroom Interaction: A Research Review* (1997).

International Library of Psychology

Editorial adviser,
Developmental psychology:
Peter K. Smith
University of Sheffield

Conceptual structure in childhood and adolescence

The case of everyday physics

Christine J. Howe

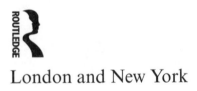

London and New York

First published 1998 by Routledge
11 New Fetter Lane, London EC4P 4EE

Simultaneously published in the USA and Canada
by Routledge
29 West 35th Street, New York, NY 10001

Typeset in Times by Routledge
Printed and bound in Great Britain by Biddles Ltd,
Guildford and King's Lynn

British Library Cataloguing in Publication Data
A catalogue record for this book is available from the British Library

Library of Congress Cataloguing in Publication Data
A catalogue record for this book has been requested

ISBN 0-415-14729-8

To the children and staff of Balfron Primary School
1987–1997

Contents

Illustrations

FIGURES

Preface

This book is an attempt to address some fundamental questions about human cognition, questions which are of relevance to both psychological theory and educational practice. The book came about, however, because of literature concerned with a widely acknowledged social problem, why a disproportionate number of pupils abandon physics. A myriad of solutions has been proposed but over the past twenty years increasing attention has been paid to the potentially subversive influence of prior knowledge. In particular, it has been argued that pupils come to physics teaching with preformed ideas about the phenomena they will be studying. These ideas undermine the formal message of teaching, resulting in failure, disenchantment and eventual abandonment. Inspired by this line of reasoning, attempts have been made to chart the preformed ideas through systematic research and to demonstrate their intrusion into the physics classroom. Thus, a literature has emerged which is focused on what is commonly referred to as 'everyday physics'. Since, despite the best efforts of national curricula and so forth, physics is seldom taught before the teenage years, this literature focuses on the thinking and reasoning of relatively senior pupils.

I became aware of the literature about eight years ago. However, reviewing it with the eyes of a psychologist and not a maker of social policy, I felt that the 'alternativeness' of everyday physics was being overplayed. Certainly, everyday physics was sufficiently unorthodox to have the dire classroom implications being claimed for it. Nevertheless, there were still distinct points of contact with the received wisdom of science, contact at the levels of both conceptual content and conceptual structure. It occurred to me that these points of contact were exactly what would be expected if, as some influential theorists have claimed, human cognition is organised around causal mechanisms. I could also see that if these theorists were correct, the implications for educational intervention would be both transparent and relatively straightforward. Despite this, I was hesitant. It was not clear to me that the points of contact applied across everyday physics. Moreover, even if they did, I was unsure whether anything conclusive could be said given data which were derived from the teenage group or older. It seemed to me that to make definite statements about the organisation of cognition and hence to draw implications for education, it would be necessary to work with younger pupils. Furthermore, this would be the case even if physics teaching continues to be directed at the older age groups.

I resolved therefore to find out everything that I could about everyday physics in the 5 to 15 age group. In principle, I could have gone even younger but methodological problems seemed to preclude this in practice. Working then with 5- to 15-year-olds, I attempted: (a) to synthesise existing studies, noting that in most cases the studies were conducted for purposes different from mine; (b) to re-analyse datasets which I had in my possession, even though once more these datasets had been obtained for different purposes; and (c) to conduct a new investigation which covered a range of physics topic areas and which had the focal issues firmly in mind. This book reports my results. I do not pretend that the results are conclusive: my three sources of information were insufficient to answer all of the relevant questions. Nevertheless, they were adequate to convince me that far from being organised around causal mechanisms, human cognition is in fact grounded in sensori-motor action and elaborated via cultural-symbolic practice. The purpose of the book is first and foremost to explain why I have reached this conclusion. However, because I am unwilling to reject one theory without having an alternative to propose, the purpose is also to sketch a model of action–symbol co-ordination which fits the data and is worth researching further, and to outline the implications of the new model for educational practice.

I have, then, a twofold aim: to clarify psychological theory and to serve educational practice. As such, I have two distinct readerships in mind, and I am acutely aware of potential tensions. Psychologists, I fear, are well represented amongst the 'disproportionate number of pupils' who have abandoned physics. Thus, they may feel uneasy about the prospect of physics topic areas, suspecting that they lack the background knowledge to make sense of children's ideas. Anticipating such feelings, I have tried to provide all the relevant physics at some point in the text. Moreover, I have restricted myself to only that part of physics which is absolutely necessary, and I have covered the material at the simplest level possible. Educationalists, by contrast, may fear that a psychological text which reports new empirical work will be burdened down with obscure statistical analyses. Although some state-of-the-art techniques could have made my text more elegant, none were essential. Thus, I have been able to restrict myself to analyses which in all cases are straightforward and which in all but one case rely on standard and widely known techniques. The exception is carefully explained. In short then, the book is intended as a cross-disciplinary venture, and I have tried hard to be sensitive to what this implies. Whether I have succeeded or not remains to be seen.

Acknowledgements

The bulk of the research for this book was conducted during the year when I was in receipt of a Nuffield Personal Research Fellowship. I am therefore greatly indebted to the Nuffield Foundation for their support. The research included a literature review which was expedited by a visit to the Children's Learning in Science Project at Leeds University. I am very grateful to the staff at Leeds for their help, and particularly to Rosalind Driver and John Leach. The research also involved re-analyses of existing datasets, primarily ones obtained by virtue of grants from the Economic and Social Research Council (C00232426) and the Leverhulme Trust (S903274). I should like to thank both organisations for their support.

The book was drafted a couple of years after the Nuffield Fellowship when I was awarded an eight-month period of study leave by Strathclyde University. I owe an enormous debt to the Department of Psychology and to the Faculty of Arts and Social Sciences for allowing me this leave. I am especially grateful to colleagues in Psychology for covering my routine duties during my leave. Many of these colleagues also participated directly in the research relating to the book as co-workers on the relevant projects, and here I should like to make special mention of Tony Anderson, David Best, Karen Greer, Jenny Low, Mhairi Mackenzie, Terry Mayes, Cathy Rodgers, Pam Smith and Andy Tolmie. Andy Tolmie in particular has been involved throughout, working on all but one of the studies and commenting in detail on a draft manuscript. I could not have completed the work without his help and I am therefore very much in his debt. Also to be thanked for his comments on a draft is Series Editor Peter Smith. Without doubt, the text is much improved by virtue of his experience and his wisdom.

Finally, I should like to thank the people who have had to put up with me while the book was in preparation. My secretary Jean Cuthill has cheerfully tolerated the numerous revisions and has done everything in her power to produce new drafts to my (probably unreasonable) deadlines. My family, my husband Willie and my children Miriam and Jeremy, have cheerfully accepted the endless hours when I have been cloistered in my study with red ink, correcting fluid and the ever-growing manuscript. It must be hard for them to accept that any book could be worth the trouble, let alone something academic. Nevertheless, they have given me unflinching support. I am very lucky and very, very grateful.

Part I Introduction

1 Everyday physics and conceptual structure

This book is intended as a contribution to cognitive psychology and educational practice. Nevertheless, it would not have been written had it not been for a simple fact of modern life, that many students experience school physics as extremely unpleasant. Indeed, many students regard school physics as a painful ordeal whose only saving grace is that it can be quickly jettisoned in favour of the humanities or the biological sciences. Thus, even amongst A-level candidates, who are themselves a selective sample, only about one-sixth of students are currently enrolled in physics.[1] This rejection of school physics led to the book because of the discussion about what lies behind it, discussion in other words of why physics proves such a nightmare that so many students wish to abandon it at the first opportunity. The discussion has been wide-ranging for there is considerable anxiety about our nation's competitiveness when such a central science is being shunned, and over the years a range of proposals have been made. One favourite is poor teaching. It is argued that when physicists are so rare and so valuable, the good and inspirational ones are unlikely to be attracted to a low-paid profession like teaching. Another is to call on the nature of physics. It is said to be too mathematical or, with phrases like 'a massless rope strung over a frictionless pulley', too abstract.

Such claims may or may not have relevance. However no matter what their truth, they have been supplemented in recent years by a different approach, and this is what triggered the book. Instead of focusing on the teaching or the subject matter, the new approach draws attention to the students themselves. It is centred on the proposal that when students embark on physics, they are not 'blank slates' with respect to the phenomena they will be studying. Rather, they are holders of strong preformed ideas which, being at variance with the received wisdom of science, lead to faulty representations during problem solving and hence to failure and eventual frustration. These preformed ideas are frequently referred to as 'everyday physics'. To support the approach, there has been a stream of studies charting the behaviour of novice students of physics while they work on typical problems, and there is little doubt that these studies do attest to preformed ideas which are deeply engrained and educationally subversive.

However, while this must be recognised, it does not necessarily mean that everyday physics is the extreme polar opposite of received science wisdom. On the contrary, points of contact are not simply possible but can in fact be readily observed. It is this paradoxical combination of similarity within difference which renders everyday physics psychologically and educationally interesting and which prompted the book.

This chapter and its successor will set the scene for what is to follow in the book's main body by explaining why everyday physics is significant from the psychological and educational perspectives. To begin, this chapter will summarise a sample of studies with novice students of physics which leaves few grounds for doubting that preformed ideas do play a role in problem solving and that the consequence of this often turns out to be problem-solving failure. The chapter will then show how, despite this, preformed ideas are not in all respects 'at variance' with received science wisdom, by virtue in fact of identifying three points of contact between everyday and received ideas. The first is that the everyday system often makes reference to variables, some of which are scientifically relevant. In other words, relations of the 'If Condition C_i then Event E_i' form are used and in some cases the conditions are not too wide of the mark. For example, many novice physicists believe that if objects are metal, they will heat up relatively quickly. This belief is in fact correct. The second point of contact is that the everyday system often calls upon causal mechanisms, and these mechanisms can also contain elements of truth. For instance, a downwards force akin to gravity is frequently recognised, even if this force is taken to operate in an unorthodox fashion. The final point of contact is that the posited relation between variables and mechanisms is in some cases suggestive of theorising. This is to say that, as with a theory, the mechanisms play a generative role in the selection of variables. Thus, it is because heat works in a particular way that metalness is seen as significant to the rate of heating. It is because of the workings of gravity that certain variables are seen as significant to resting or falling.

Having identified these points of contact, the chapter will then begin the task of explaining why they give everyday physics its great significance. Its first step here will be to show how the points of contact concur exactly with the predictions of a recent and influential approach within cognitive psychology. The approach centres on the claim that the generative power of mechanisms is no more and no less than a 'primitive' of human cognition, with theoretical structure being as a consequence an entrenched feature from early in life. This being the case, there is a strong expectation that theoretical structure will be identifiable in mechanism–variable relations, meaning that the third point of contact is supportive evidence. In addition though, understanding mechanisms will clearly, on this approach, have the force of a cognitive imperative and this imperative will also operate from early in life. However, given the structuring of physics education in the industrialised world, novice students are typically teenage or older. Thus, they will have had plenty of time to respond to the imperative, and should have attained a fair understanding. As a result, there is an expectation of some orthodoxy over mechanisms, meaning that the second point of contact also supports the approach.

In addition though, to the extent that the approach is endorsed, so the teaching problems engendered by everyday physics become less severe than they superficially seem. After all, if variables are generated by mechanisms, the implication is that teachers should focus their attention upon the latter. The links and contrasts between the mechanisms of everyday and professional physics should be mapped out carefully, and strategies should be developed for fostering orthodoxy. If this were done, orthodoxy over variables should fall out naturally. In addition though, the partial adequacy of mechanisms means that fostering adequacy may not prove particularly difficult. Thus, if the approach just outlined is correct, there are also strong and positive implications for educational practice, meaning that the findings from everyday physics are beginning to seem like very good news indeed.

However, is this sense of endorsement really justified? For one thing, does the approach to cognitive psychology *require* empirical support from everyday physics? Has it not become established already with reference to other evidence or perhaps to logical necessity? Moreover, even if further evidence is required, can the findings from everyday physics be regarded as conclusive? The present chapter will end by raising these questions, to see them discussed further in Chapter 2. The answers across the two chapters will be 'Yes, empirical evidence is required', but 'No, the evidence from everyday physics is not conclusive'. Rather, the evidence has established everyday physics as a key arena for further research. Indeed, the required research should not simply bear incisively upon the aforementioned approach and its educational ramifications; it should also resolve core issues about cognition in general. As Chapter 2 will make clear, the research in question will be developmental, tracing changes with age with regard to variables, mechanisms and their interrelation. This then is where the present chapter is heading, towards the acceptance that the similarities yet differences between everyday physics and science orthodoxy make the former an arena for developmental research of some significance. The results of such research will occupy us from Chapter 3 onwards.

THE 'ALTERNATIVENESS' OF EVERYDAY PHYSICS

The emphasis will, then, be on the development of the preformed ideas that are eventually brought to physics, the development in other words of an everyday physics. However, to put the enterprise in context, we need, as signalled already, to look at everyday physics at the point of formal instruction. Does it really show the 'alternativeness' which many have claimed, and yet does it also show the points of contact which render it significant? To answer the question, the present section will summarise a sample of the studies mentioned already. These are studies which take school pupils or college students who have recently embarked on the study of physics and chart the strategies which they use while working through a characteristic series of problems. For clarity, the section will organise the studies into two subsets: those concerned with predictive problem solving and those concerned with explanatory. Having reviewed the studies, the section will make comparisons with science orthodoxy. Is everyday physics an alternative and subversive entity which nevertheless shares some properties with received ideas? Moreover, what does this paradox signal for theory and practice?

Predictive problem solving

A favoured approach to physics teaching is to present students with constellations of variables and ask them to predict outcomes from given values. Empirical problems of this kind are used, as for example when students are asked to predict the temperature loss per unit time of water which is presented to them in containers of varying material. However, theoretical equivalents are also popular when for example students are told that an object falls from a particular height, and asked to calculate its speed on landing. Many research projects have assessed students' success and failure on these kinds of problems. Indeed, there have been cross-nation surveys to this effect. However in their own right, such projects have little to say about the existence of everyday physics. Whatever else is involved, obtaining the correct answer is at least partly a function of computational skill. Thus, by simply looking at success or failure, it is impossible to differentiate the effects of preformed ideas from the effects of mathematics.

Greater insight might be obtained by looking at the general direction of solutions rather than bothering about their accuracy in detail. Thus, the issue in our empirical example would be whether greater temperature loss is predicted in, say, a metal container than a polystyrene one. Whether the absolute value was correct or not would be beside the point. The issue in the theoretical example would be whether landing speed is predicted to be greater than, equal to or less than starting speed, again regardless of computational accuracy. However while this more global analysis would undoubtedly be preferable, there is still potential ambiguity about the inferences to draw. Guessed solutions would, in some circumstances, be hard to differentiate from those motivated by preformed ideas. Thus, an even better method would be to combine global analysis with students' accounts of what their predictions are based on. In cognitive science, the traditional approach to obtaining accounts of any problem-solving activity is to ask students to 'think aloud' while performing the task. However, an alternative approach which is equally immediate and surely more natural is to question students directly as to why they responded in the way they did.

Taking predictions plus follow-up questioning as the preferred approach, there are a number of relevant studies. However, although these studies have covered a range of topic areas, two themes recur and thus provide particularly convincing evidence on the issues at stake. The first theme is object fall after horizontal motion, as for example when a ball rolls off a cliff or an apple core is tossed from a moving car. In these circumstances, the object will fall following a parabolic path in the direction of the horizontal motion. This results from the interplay of the progressive deceleration in the horizontal direction and, due to the force of gravity, the progressive acceleration in the vertical. It is, importantly, nothing to do with the dissipation of a horizontal force, for there are no forces in that direction subsequent to the object being set in motion. The question that most of the studies have addressed is whether students appreciate this point.

The first, and best known, of the studies was conducted by Michael McCloskey and various colleagues and is summarised in McCloskey (1983a). The study was part of an ambitious programme of research, utilising a range of problems within the basic paradigm, for instance metal balls dropped from an aeroplane or, as in Figure 1.1, sliding over a cliff. In much of the research, students were simply asked to predict the paths that the objects would follow. This established that forwards parabola are correctly anticipated in fewer than 50 per cent of the cases, with the errors including backwards parabola, vertical straight lines, diagonal straight lines and horizontal straight lines followed by downwards paths of varying shapes. The study of interest also involved prediction and replicated the basic results. However in this study, the prediction phase was followed by interviews where students were asked to explain their responses. Thirteen students were interviewed, all undergraduates with limited expertise in physics. Eleven showed strong commitment to a gradually dissipating horizontal force, indicative as McCloskey points out to a concept of 'impetus'.

Similar results were obtained by Aguirre (1988) in a study with 15- to 17- year-old pupils. Aguirre's apparatus was a large flat surface positioned at an angle to the floor. There was a plunger in the top left hand corner which could propel a plastic block onto the surface. With horizontal velocity under the influence of gravity, the block's path would be parabolic. However as with McCloskey's study, the pupils seldom appreciated this, making a similar array of inaccurate predictions. Moreover, in justifying their predictions, the pupils frequently cited a horizontal force akin to impetus. Other studies, for example Whitaker (1983), produce similar results, but there are also subtle differences. One appears in the work of Eckstein and Shemesh (1989) on descent from a moving vehicle, in this case a cart. Adult and child novices in physics were asked whether (and why) a ball falling from a pole attached to the cart would land in a cup placed directly below. Two groups were identified in terms of response. One group answered incorrectly that the ball would miss the cup, usually calling on impetus. The other group by contrast answered correctly that the ball would land in the cup but justified this with reference to a quasi-magnetic relation between ball and cart. As

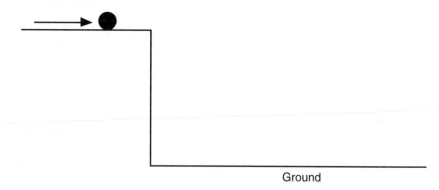

Figure 1.1 An example of the problems used by McCloskey (1983a)

The diagram shows a side view of a cliff. The top of the cliff is frictionless (in other words, perfectly smooth). A metal ball is sliding along the top of the cliff at a constant speed of 50 miles per hour. Draw the path the ball will follow after it goes over the edge of the cliff. Ignore air resistance.

one respondent put it 'The ball is like one unit with the cart and the post. It's all one body'(Eckstein and Shemesh, 1989: 330).

The 'one-ness' reported by Eckstein and Shemesh is not the same as the impetus reported by McCloskey and Aguirre. Nevertheless, they both amount to data which are grist for our mill, for they both bear witness to ideas that are unlikely to have come from orthodox science. They do however relate to a single theme. Thus, it is as well that, as mentioned earlier, there is parallel and equally voluminous research on a different theme. The theme is electricity, and in particular the consequences of differing arrangements of resistors and batteries. To appreciate the issues, consider Figure 1.2 which depicts some possible, but very simple, electrical circuits. Imagine for the moment that X refers to a battery, Y, Y' and Y" to identical resistors, and Z to an indicator of current (perhaps nothing more than a light bulb). In these circumstances, (a) has one resistor while (b) and (c) have two connected in series. In both cases, the consequence would be to decrease the current relative to (a). Although (d) also has two resistors, they are connected in parallel. The consequence here would be to increase the current relative to (a). If by contrast X referred to a resistor and Y, Y' and Y" to identical batteries, the consequence of (b) and (c) would be to increase the current relative to (a). The consequence of (d) would be to maintain the current of (a).

A number of studies have used circuits along the lines of Figure 1.2 to explore everyday physics. One such study was reported by Gentner and Gentner (1983). Here thirty-six high school and college students 'screened to be fairly naive about physical science' (Gentner and Gentner, 1983: 117) were asked to predict the current in circuits like (a), (b) and (d), with doubling of the resistors in some problems and doubling of the batteries in others. The students were also quizzed about their 'mental models' as to

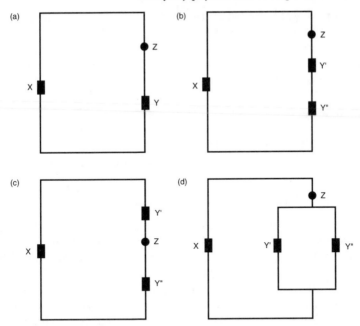

Figure 1.2 Some simple electrical circuits

how electricity travels. Gentner and Gentner identified two main models labelled, in a fairly self-explanatory fashion, the 'water flow' and the 'moving crowd'. They hypothesised that students who subscribed to the water flow model would perform well when the batteries were doubled. This is because the serial set up, that is (b), should remind them of tanks at different heights, height being the sole consideration relevant to water pressure. The parallel set up, that is (d), should remind them of tanks at the same height where water pressure is therefore identical. Gentner and Gentner further hypothesised that students who subscribed to the moving crowd model should perform well when the resistors were doubled. This is because the serial set up should remind them of how a sequence of turnstiles serves to slow crowds down. The parallel set up should remind them of how a choice of turnstiles serves to speed them up. Gentner and Gentner's results provide strong support for both hypotheses.

Related to Gentner and Gentner's research is a study reported by Shipstone (1985). In this study, a paper-and-pencil test was administered to pupils aged 11 to 18 at three British comprehensive schools and to A-level students at a sixth form college. All participants had embarked on the study of electricity. One item in the test used a circuit like (c) in Figure 1.2, indicating clearly the direction in which the current was flowing and interpreting the doubled symbols as resistors. Using a multiple choice format, pupils were asked to predict the consequences of increasing and decreasing the resistance before and after the indicator. (If the current was flowing clockwise in (c), Y' would be before the indicator and Y" would be after.) The pupils were also asked to justify their responses in writing. In reality the location of the resistors makes no difference; it is their strength that counts. However, a very large number of pupils failed to appreciate this, believing that only resistors before the indicator had any significance. As one pupil put it 'R_1 [as it was labelled in Shipstone's study] is after the lamp . . . hence it will not hinder the voltage' (Shipstone, 1985: 41). Interestingly, the frequency of this belief increased from 30 per cent of the pupils at age 11 to 12 to 80 per cent at age 13 to 14. Moreover, although its frequency then fell away again, it was still proposed by 30 per cent of the sixth form sample.

Explanatory problem solving

The studies just described provide strong evidence for the existence of preformed ideas. Nevertheless, they are limited to one form of physics problem solving, namely prediction. Thus, based on these studies alone, it would be difficult to claim that interference from preformed ideas is an entirely ubiquitous phenomenon. In particular, problem solving in physics classrooms is as likely to involve the explanation of events that have already taken place as it is to involve their prediction. Indeed, as with prediction, it is possible to think of empirical and theoretical instances of explanatory problem solving. Empirical instances are easy to come by, for they occur after every laboratory experiment or demonstration. They occur, for instance, when pupils bring tap water to the boil on their Bunsen burners and are asked to explain why the temperature remains constant at 100°C. They occur also when the teacher encases a burning candle in a bell jar and demands to know why its flame goes out. However, theoretical instances are equally common, particularly in standard texts. Two intriguing examples from Halliday and Resnick (1988) are 'Could you weigh yourself on a scale whose maximum reading is less than your weight. If so, how?' (Halliday and Resnick, 1988: 97) and 'You must have noticed (Einstein did) that when you stir a cup of tea, the floating tea leaves collect at the centre of the cup rather than at the outer rim. Can you explain this (Einstein could)?' (Halliday and Resnick, 1988: 118).

No matter whether the subject matter is empirical or theoretical, there is a subtle difference between explanatory problem solving and predictive. With explanatory problems, the conceptual base must necessarily be articulated as part of the solution. Thus, in contrast to predictive problems, there is no need to ask students to justify their solutions to get at their underlying ideas. If such ideas exist, they should be revealed in the solutions themselves. As Scriven (1962) has shown, this difference between explanation and prediction has major implications for the philosophy of science. For us, it has the more mundane methodological implication of allowing evidence relevant to everyday physics to be obtained during ordinary classroom activities without needing follow-up questions, a point that has not gone unnoticed by researchers in the field. One group of researchers has focused on the deceptively simple phenomenon of being at rest, and their data give every impression of being collected in natural contexts.

For instance, Minstrell (1982) describes a study that 'was conducted entirely in the natural setting of the physics classroom' (Minstrell, 1982: 10), via in fact the tape recording of discussions and the careful scrutiny of homework papers and classroom tests. The subjects were American high school students, and their task was to explain what keeps a stationary book at rest on a table. The received view, derived from Newton's Third Law of Motion, is that gravity and the table exert equal but opposite forces on the object. However, while virtually all students recognised the operation of gravity, only half were aware of the opposing force of the table. Moreover, even the students who were aware did not always realise that the opposing forces were equal: many were convinced that the downwards force must exceed the upwards. In addition, whatever their views regarding gravity and the table, many students believed that wind or air pressure were also playing a part. Finally, gravity was seen as a variable phenomenon, often thought to disappear at ground level. Thus, had the book been on the floor rather than on a table, the responses would have differed.

Findings reminiscent of Minstrell's emerged from a study reported by Gunstone and White (1980, 1981). One component of this study involved placing a blackboard duster on a book held about two metres above a bench and asking a sample of students (175 in total) to write down why the duster failed to move. This time, the students were university undergraduates and thus, unsurprisingly, the proportion of errors was lower. Nevertheless, sixteen students indicated that the book did not exert a force on the duster, while five stated that the book's force was not equal to the force due to gravity. Moreover, a slightly more complex arrangement, presented this time to 463 students, revealed a new set of confusions. This arrangement involved a bicycle wheel 'pulley' supporting a bucket of sand at one end and a block of wood at the other. The bucket was markedly higher than the block although the system was stationary. Because the system was stationary, it would be evident to an expert physicist that the bucket and the block were equal in weight, but 122 students reasoned to the contrary. They argued along the lines of 'the block is heavier than the bucket; since the block is nearer the floor, hence it must be heavier' (Gunstone and White, 1980: 38). This confusion of height and weight is interesting in the context of Minstrell's research. There, it will be remembered, some students thought gravity dissipates as the ground is approached. However in conventional physics, weight is defined as mass × gravity, and Gunstone and White find weight being thought to increase on approaching ground level. This suggests highly unorthodox beliefs about the relation between weight and gravity, a point that will be taken up again in subsequent chapters.

The scenarios used by Minstrell and Gunstone and White are deceptively ordinary, and the same applies to a contrasting body of research also concerned with explanation in physics. This research was stimulated by

Brook *et al.*'s (1984) survey of secondary school pupils' understanding of heat. A group of 15-year-olds was the focus of the survey, but some items were presented to 11- and 13-year-olds. Amongst the items were ones asking about the contrasting 'feel' of objects despite constant ambient temperature. One such item invited the pupils to explain why the metal and plastic parts of bicycle handlebars feel different on a frosty day. Relatively few pupils gave an adequate explanation, namely that the metal parts conduct the body's heat more rapidly than the plastic. However while Brook *et al.*'s study demonstrates this point, it does not attempt an analysis of the inadequate responses that the pupils gave. Thus, it leaves the nature of pre-existing ideas completely unclear.

Recognising this, Clough and Driver (1985a) attempted a more probing study involving eighty-four 12- to 16-year-old pupils. The pupils were interviewed individually regarding three problems. For the first problem, a metal spoon, a pottery spoon, a wooden spoon and a plastic spoon were dipped into a mug of hot water, and the pupils were asked to explain why the metal spoon felt the hottest and the wooden and plastic spoons the coldest. For the second problem, the pupils were asked to explain why a set of metal plates felt colder than a set of plastic plates when both sets had been left in the same room overnight. The third problem was Brook *et al.*'s handlebar item, based in this case on a drawing. There was quite a lot of variability between the problems as regards the pupils' responses. Nevertheless, with each problem, a sizeable proportion talked in terms of the propensity of metals to let heat in, let heat out and/or let cold in. Another sizeable proportion invoked further properties of the objects, for example thickness, colour or smoothness. Finally, 19 per cent of the responses to the first problem, 45 per cent of the responses to the second and 40 per cent of the responses to the third were 'mixed' or 'uncodeable', attesting to considerable cross-pupil variability.

Everyday physics and science orthodoxy

We have now considered two types of problem solving: predictive and explanatory. Within each type, we have considered two kinds of problem: problems relating to object fall and electricity for prediction, and problems relating to the 'at rest' condition and heat transfer for explanation. In every case, there is compelling evidence that preformed ideas about the topic area influence the proposed solutions. Very occasionally, the ideas lead to solutions which are partially correct. However, totally correct solutions are virtually unheard of, and even partial correctness is seldom achieved across a range of problems. Thus, the everyday system is not simply real; it also exerts an adverse influence on problem solving. As such, it may indeed be relevant in the sense signalled at the start of the chapter, as a contributory factor to the mass exodus from school physics. Recognising this, there is a tendency for those concerned with such matters to characterise everyday physics in terms which emphasise its deviant nature. Thus, we have 'misconceived ideas', 'alternative frameworks' and 'conflicting theories'. Everyday physics has even been likened to a competing 'paradigm' in the sense of Kuhn (1962). Following such images, teaching gets viewed in confrontational terms as a struggle between acceptable and insurrectional ideas.

Without doubt, everyday physics creates problems which teachers have to address. However, this does not mean that everyday physics is the polar opposite of science orthodoxy and careful reading of the material just discussed will, I think, bring any assumptions along these lines sharply into question. Embedded in the material are several points of contact between everyday and received ideas which must not be lost sight of. In the first place, everyday physics clearly makes reference to *variables*, which are akin in some respects (if not many) to 'laws of nature'. Predictive problems are typically expressed in terms of variables. Thus, the fact that everyday ideas can be co-ordinated with these problems is evidence in its own right for reliance on variables. More specifically though, Gentner and Gentner's (1983) work on electricity demonstrates the significance of the variables 'number of batteries', 'number of resistors', 'parallelism (vs. seriation) of batteries' and 'parallelism of resistors'. Shipstone's (1985) work achieves a similar point for the variable 'location of resistors'. Clough and Driver's (1985a) work on heat transfer demonstrates the significance of the variables 'metalness', 'thickness', 'colour' and 'smoothness'.

Strangely, there is one attempt in the literature to deny variable-based representation, a paper by Yates *et al.* (1988). This paper is offered as a critique of McCloskey (1983a), arguing that the results reported there are more compatible with the matching of events to situation prototypes. These prototypes are enacted mentally to produce the predictions. Since the notion of a situation prototype implies the holistic processing of events, Yates *et al.* are in explicit opposition to the use of variables. They even report a study to support their argument. This study was concerned with the effects of presenting two McCloskey-style problems in conditions that should (and probably did) manipulate the participants' focus of attention. Predictions varied greatly as a function of presentation condition, bearing witness to considerable situation specificity. However, while situation specificity is predicted given a prototype representation, it is not precluded by the use of variables. It can, quite simply, be achieved by assigning different values to the variables, a point also made by Springer (1990). Interestingly, although Yates (1990) has responded roundly to Springer's critique, this is one issue that he seems to shirk.

This is not, of course, to argue that situation prototypes are never used. Working also with McCloskey-style problems, Kaiser, Jonides and Alexander (1986) have evidence for reference to prototypes in conditions of extreme familiarity.[2] The point is that Yates *et al.*'s evidence against variables is weak, which is not surprising since as we have seen already variables are manifestly used. Indeed, the variables called upon in the studies of electricity and heat transfer were not simply unmistakable. They were also in some respects scientifically relevant. The number and parallelism of batteries and resistors are relevant to electricity, even if the operation of these variables is somewhat different from what Gentner and Gentner's students supposed. Metalness, thickness and colour are relevant to heat transfer and, with the first two at least, in precisely the fashion that Clough and Driver's pupils outlined. Thus, for this reason as well as the simple fact of variables, we have, I think, grounds for hesitating before dismissing everyday ideas as irretrievably 'alternative'.

Within professional science, there are instances where laws of nature stand on their own. However, there are at least as many instances where laws and the variables they call upon, attain their significance by being embedded in causal mechanisms. Thus, it is a second point of contact between everyday and received ideas that *causal mechanisms* abounded in the research just considered. Causal mechanisms were apparent in McCloskey's (1983a), Whitaker's (1983) and Aguirre's (1988) 'impetus', Eckstein and Shemesh's (1989) 'quasi-magnetism', Gentner and Gentner's (1983) two models of current flow, Minstrell's (1982) air/wind 'pressure' and 'variable gravity', and Clough and Driver's (1985a) 'heat' and, interestingly separated, 'cold'. As we have seen already, some of these mechanisms are way off beam by the standards of professional science. However, this is not true of them all. Although gravity is not, in reality, variable in the way Minstrell's students presumed, it was, nevertheless, a significant factor in the situation that he set up. Likewise although heat does not 'flow' in the sense of Clough and Driver's pupils, it remains the genuine causative agent in effecting temperature change.

Of course, the fact that causal mechanisms are acknowledged within everyday physics does not necessarily mean that they are utilised in a fashion that is consistent with science. There is, in particular, the issue of whether mechanisms provide contexts from which variables take their meaning, whether in other words variables are embedded in a mechanistic base. Scrutinising the studies discussed so far, it has to be recognised that most consider mechanisms or variables but not both together. However, there are two exceptions and both are encouraging. The first is the Gentner and Gentner study where the mechanisms (the models of current flow) dictated how the variables operated. The second is the Clough and Driver study where the images of flowing heat and cold dictated reference to metalness as a significant factor. Thus, in both cases, the mechanisms were pivotal and could even be said to be 'generative' in variable selection.

For Wellman (1990), this generativity would be of the greatest importance, for it is the crux of what he regards as the genuinely theoretical. Theories, according to Wellman, are centred on causal explanatory mechanisms, and I think that he is probably right. Although philosophers of science have debated the concept interminably, they seem to agree that, no matter what other characteristics theories may have, they are indisputably representations that revolve around mechanisms. If this gloss is correct, we should be justified from the evidence just discussed in treating everyday physics as having a theoretical dimension. Our evidence relates to electricity and heat transfer, but others have come to the same conclusion with different topic areas. For example McCloskey (1983b) claims that 'it is therefore the misconceptions embodied in an intuitive physical *theory* that occasionally give rise to errors in judgement about motion. The intuitive theory bears a striking resemblance to the pre-Newtonian theory of impetus' (McCloskey, 1983b: 114A). A similar line is taken in Gunstone and Watts' (1985) declaration that 'some of us who are exploring these issues have described students' conceptions of force and motion as Aristotelian, others have described the conceptions as similar to the mediaeval impetus *theory*' (Gunstone and Watts, 1985: 88).

McCloskey and Gunstone and Watts are writing from a science education perspective. Thus, if a note of enthusiasm can be detected in their claims, it is probably because of the positive educational implications which, as noted earlier, theorising might be taken to carry. To recap, if mechanisms generate variables as theoretical structure implies, there may be no need educationally speaking to do anything about variables. If the mechanisms of everyday physics are properly understood and strategies are devised for removing their inadequacies, the problems with variables may take care of themselves. Moreover since the inadequacies in mechanisms are not necessarily overwhelming, strategies for removing them may not be hard to find. All in all then, the pointers are towards a teaching strategy which is of wide applicability and entirely straightforward. No wonder it has appeal in educational circles.

However, as noted earlier, educationalists are not the only individuals likely to obtain satisfaction from what has preceded. The combination of theoretical structure and semi-adequate mechanisms is exactly what, given novice students of physics, a group of cognitive psychologists would expect to find. This is because these psychologists have represented theorising as a primitive of human cognition, and hence treated mechanisms as crying out to be understood. By the age at which physics is typically taught, that is teenage or older, responses to that cry should have taken understanding beyond the rudimentary, making the semi-adequacy of mechanisms consistent evidence. However, the fact that the evidence is consistent does not make it either necessary or sufficient to prove the point, and this is what we need to consider next for it bears crucially on what has been described so far. Accordingly, the next section will address the need for the evidence, by outlining in detail how theorising obtained its status as a primitive and whether that status is currently unassailable quite apart from everyday physics. Issues relating to sufficiency will be discussed in Chapter 2.

THE CONSTRAINING OF HUMAN COGNITION

The idea that theorising and hence mechanisms are primitives stems from attempts within cognitive psychology to solve the classic problem of induction.

The problem, expressed crudely and simply, is that there are an infinite number of ways in which reality can be segmented. Nevertheless, we do not merely move, often with minimal reflection, to some segmentations rather than others. We also agree with our fellow human beings over the segmentations that we prefer. For example, we all see cats vs. dogs and good vs. bad insulators as acceptable segmentations, and cats plus poodles vs. other dogs, and good insulators plus metal vs. other poor insulators, as unacceptable. The question of how segmentation becomes constrained has been asked on many occasions, and many answers have been given. However, recently cognitive psychologists have been turning to causal mechanisms, and this is the basis for the latter's privileged status. Recognising this, the present section will review the arguments which have led to mechanisms being proposed as the solution to the problem of induction. It will accept that given the conceptualisation of the problem which the relevant psychologists have adopted the arguments look compelling. Nevertheless, they do not constitute a proof, opening the way for empirical investigation and perhaps for everyday physics.

Mechanisms as conceptual constraints

A lucid and therefore influential attempt to use causal mechanisms to solve the problem of induction appears in the cognitive psychology of Murphy and Medin (1985, but see also Wattenmacher *et al.* 1988). Murphy and Medin express the problem in terms of why some groupings of phenomena 'are informative, useful and efficient, whereas others are vague, absurd or useless' (Murphy and Medin, 1985: 289). Murphy and Medin's solution is that 'representations of concepts are best thought of as theoretical knowledge or, at least, as embedded in knowledge that embodies a theory about the world' (Murphy and Medin, 1985: 289). The term 'theory' is explicitly acknowledged to connote 'a complex set of relations between concepts, usually with a causal basis' (Murphy and Medin, 1985: 291) and 'a network formed by causal and explanatory links' (Murphy and Medin, 1985: 289).

To reach their solution, Murphy and Medin discuss and reject the possibility that segmentation is legitimated with reference to perceptual similarity. Perceptual similarity raises the question of why similarity on some dimensions does not produce conceptual equivalence, for instance why cats and dogs are not treated as equivalent despite tails, four legs and fur. The answer can, Murphy and Medin acknowledge, be partly found in the human perceptual apparatus, which 'selects' certain features and discounts others. However, this is not a complete solution, for it says nothing about conceptual equivalence under perceptual difference. Murphy and Medin's example here is the Jewish concept of clean and unclean animals, the former including gazelles, frogs and grasshoppers and the latter camels, mice and sharks, and thus neither showing marked perceptual coherence. In reality, we do not have to move so far from the science classroom to make the same point. Gelman and Markman (1986) presented children with pictures showing three living creatures. Two of the creatures were biologically related, for instance a blackbird and a flamingo, and two were perceptually similar, for instance a blackbird and a black bat. Biological properties were identified for two creatures in each picture, for example 'This bird (the flamingo) gives its baby mashed-up food' and 'This bat (the bat) gives its baby milk'. The children were then questioned as to the biological properties of the third creature, for example, 'Does this bird (the blackbird) give its baby mashed up food or milk?' The children consistently responded in terms of biological relationship and not perceptual similarity.

To reinforce the inadequacy of perceptual features, Murphy and Medin cite well-known research by Chapman and Chapman (1967, 1969). Here trained psychotherapists, and indeed untrained subjects, detected correlations between psychometric test results and psychological disorders when in fact there were none. However as well as confirming the inadequacy of perceptually based models, the Chapman and Chapman research was seen by Murphy and Medin as providing direct evidence for the involvement of mechanisms. Their point was that subjects were seduced into 'illusory correlations' because they held theories which deemed the symptoms revealed in the tests to be caused by certain disorders. Subsequent work by Medin *et al.* (1987) has underlined this. For instance, Medin *et al.* report an experiment where subjects were asked to sort medical symptoms into categories. Their main finding was that causal linkage was a good predictor of performance. Thus, dizziness and earache which can be linked causally (by anyone with a smattering of experiential and/or medical knowledge) were more likely to be placed in the same category than were, say, sore throat and skin rash.

The point is that in these contexts of medicine-cum-psychotherapy, theory-driven linkages were preferred over the ones that would be derived from perception. This, for Murphy and Medin, became the crux. If theory-driven linkages are preferred over perceptual, then it must be theories that are determining how linkages are made. As noted, Murphy and Medin's allegiance lies in cognitive psychology. However, their arguments are echoed elsewhere, for Harré and Madden (1975) have come to similar conclusions despite working from very different beginnings. Harré and Madden's starting point is, in fact, the epistemology of David Hume, an epistemology which presupposes the segmentation of reality along perceptual lines. Hume's work has been repeatedly and soundly criticised during the two centuries of its existence. The criticisms are rehearsed by Harré and Madden, who then assert that the solution is to presume 'powerful particulars' that obtain their effects through 'the workings of generative mechanisms' (Harré and Madden, 1975: 141).

Space does not permit a detailed account of either Hume's epistemology or Harré and Madden's rebuttal. However, a taste can be obtained by considering the distinction between 'nominal' essences and 'real' ones. As defined by John Locke, nominal essences are those features of phenomena that allow us to recognise them for what they are. Thus, material, thickness and surface area contribute to the nominal essence of conductivity. Real essences are those features which warrant the designation of some essences as nominal as opposed to accidental. Thus, the ability to transmit heat energy is the real essence of conductivity, for it explains why material, thickness and surface area are crucial when newness can never (not even if all the efficient saucepans are new) be more than accidental. As we have seen in effect already, the distinction between nominal and real essences falls out naturally if we recognise causal mechanisms (or powerful particulars). However, it simply cannot be made if perception is the sole basis of knowledge. Then, of course, nominal and real essences reduce to different instances of perceptual features.

It is interesting how similar this line of reasoning is to Murphy and Medin's. As we have just seen, the collapse of the nominal vs. real distinction entails the collapse of the nominal vs. accidental. However, the collapse of the latter amounts to the removal of criteria for preferring some segmentations over others, which is exactly what Murphy and Medin were talking about. Indeed, it very explicitly opens the floodgates regarding which segmentations are made, and thus links fairly directly with the problem of induction. In this context, it is intriguing to find Harré and Madden writing, a trifle optimistically I feel, that 'it would be misleading to say that our counter-analysis of "causality" solves the problem of induction, for the rendition of the problem is essentially Humean in the first place' (Harré and Madden, 1975: 71).

Alternative constraints on segmentation

It is gratifying to witness cognitive psychologists and philosophers coming to essentially similar conclusions. However, are the conclusions established beyond all reasonable doubt? If they are, we have to accept that causal mechanisms play the strongest possible role in human cognition. Causal mechanisms must be operating whenever conceptual distinctions show signs of being preferred, whenever indeed there is cultural consensus over which distinctions are made. This means, amongst other things, from early in childhood, for very young children have been observed to associate objects in a principled and consensual fashion. For instance, Nelson (1973) presented 19- to 22-month-old children with a series of eight-object sets. One set contained aeroplanes identical apart from size, another set animals identical apart from colour. The sets were presented with the objects arranged in a haphazard fashion. Nelson found that over 70 per cent of the children respected the discriminating features when 'putting the objects the way they ought to be'. Likewise, Daehler et al. (1979) presented children aged 22, 27 and 32 months with a number of standard objects, and asked them to select from arrays the objects that went with each standard. The relations between standards and targets varied from identity through superordination (dog;

other animal) to complementarity (knife; fork). Regardless of relation, the children's ability to select the targets was above chance level.

Under the model we are considering, the data presented by Nelson and Daehler *et al.* would have to be regarded as indicative of theorising. In other words, the reason that the children sorted by size, colour, functional complementarity and so on is that they held theories which told them to do so. To their credit, both Murphy and Medin and Harré and Madden recognise that early theorising is a consequence of their model, and Harré and Madden offer an explanation as to how this could happen. They call upon the widely cited experiments of Michotte (1963). In these experiments, adult observers witnessed simple scenarios where one triangle moved towards another, with the second triangle beginning to move the instant that the first one reached it. Observers invariably reported that the first triangle caused the second triangle to move. For Harré and Madden (and indeed for Michotte), this is evidence that the operation of causal mechanisms is recognised directly. It must therefore be 'wired-in' to the human constitution, with the implication of availability at birth.

It is important to note that Harré and Madden do not refer to Michotte because they feel that the implications of their thesis for children is likely to be their undoing. On the contrary, they regard their thesis as above empirical challenge, and hence they use Michotte as a possible account of something which will definitely be accountable in some terms or the other. If Michotte proves to be inappropriate, then there must be something else. Confidence indeed, but is it well founded? At first sight, it might appear to be, for if we scrutinise the cognitive literature for constraints on induction that are alternative to mechanisms, most of the apparent candidates turn out to require mechanisms (or something equivalent) to give them explanatory value. Take for example the notion of 'scripts'. This notion has been promulgated in the literature by Schank and Abelson (for example, 1977) and refers to integrated series of routine events. Scripts are typically discussed in the context of social events, restaurant visits being a favourite. However, it is easy to imagine scripts for the physical events discussed in the previous section, scripts for the horizontal then falling motion of objects and scripts for the temperature profiles of substances in various containers.

Without doubt, scripts provide principles for segmenting reality: as Nelson (1983) has pointed out, phenomena will be seen as similar to the extent they play the same scripted roles. Nevertheless, while this is true, scripts are themselves segmentations of reality and thus also need explaining. It is in fact just as appropriate to ask why we recognise restaurants vs. shops rather than Maxim's vs. other restaurants plus shops as it is to ask our earlier question about cats and dogs. In view of this, scripts would be seen by the theorists we are considering as entailing mechanisms rather than substituting for them. Explicit acknowledgement of this comes in Wellman (1990) when he writes that 'some aspects of scripts are made sensible only by reference to and dependency on our framework theories' (Wellman, 1990: 135).

Similar points can be made about the use of 'rules' in knowledge representation. Rules are particularly familiar in the context of linguistic representation but they have recently been extended to knowledge in general by Holland *et al.* (1987). Holland *et al.* state explicitly that their work relates to everyday physics, and the research of McCloskey (1983a) is used as an example. What the work centres on is a series of so-called 'production rules'. Production rules take the form of condition–action pairs, for example 'If an object is long and slithery, regard it as a snake' and 'If an object is propelled into space, predict that it will fall diagonally downwards.' Holland *et al.* assume that cognitive activity is triggered by problem solving. Thus to the extent that the features of a current problem match the conditions in an existing rule, the rule will be activated. When the conditions in several rules are matched, each of the rules will be activated and algorithms will be applied to assign one priority. In any event, a single rule will emerge which will guide problem-solving activity. If this is successful, the rule will be strengthened and its likelihood of being assigned high priority in the future will be increased. If problem solving is unsuccessful, the strength of the rule will, at the very least, diminish. There may in addition be some changes to the rule and/or the introduction of a new rule.

Holland *et al.* are not only well aware of the problem of induction; their book is also explicitly offered as an attempt to solve it. Yet when they address the problem directly, their main source of constraints in terms of production rule change is the problem-solving context and the feedback on success. The history of linguistic representation shows only too clearly what a risky line this is. It is possible to express the grammars of natural languages using production rules (see for example Winograd, 1982). Moreover the grammars of natural languages are undoubtedly called upon to solve problems, namely the problems of conveying communicative intentions. Feedback in terms of communicative effectiveness is often given.[3] Thus, it is highly relevant that a classic paper by Gold (1967) proves that given usage and feedback alone consensual beliefs about grammar are logically impossible. Holland *et al.* do not discuss Gold's work, but they come close to acknowledging difficulties with their own approach when they tackle the question of implausible rules like 'If you see a pebble with a red stripe, then you sneeze three times'. Seeking to explain why such rules are in fact implausible, they call on 'the causal theories of any system not born yesterday' (Holland *et al.*, 1987: 81)!

The introduction of action

What we have seen with both scripts and rules is structures intended to impose constraints on segmentations of reality needing further constraints to guarantee consensus. In both instances, a plausible candidate for those constraints has been theorised mechanisms. In view of this, it seems that a powerful case has been made for viewing mechanisms as the keystone of cognition, that is as conceptual primitives around which knowledge revolves. However, is the case so powerful that it can be regarded as established on theoretical grounds alone without requiring evidence? If it can, the apparent endorsement from everyday physics might be gratifying, but it could hardly be treated as crucial. The case for mechanisms is undoubtedly powerful, but it rests on one key assumption and this undermines any claims that it might have to being a proof. Specifically, the argument throughout has been that mechanisms are required to overcome the problems that *perception* would give rise to. However, this argument has force only to the extent that cognition is perceptually based in the first place, and hence that in the absence of mechanisms perception would run riot. Perhaps though, this concedes too much to the Humean tradition. Many scholars would say that it does, arguing that cognition is in reality grounded in action and thereby muddying the waters considerably. In particular, once the possibility of action-based knowledge is conceded, the issue of mechanisms stops being theoretical as a matter of principle. Empirical resolution correspondingly becomes crucial. However what form of empirical evidence is required, and is there a role for everyday physics?

To answer the two questions posed, we need to ascertain the implications of action-based cognition for theoretical structure in general and for everyday physics in particular. This will be one of the main themes of the chapter to follow, and to prepare the way we need to summarise what has been established so far. First, evidence has been presented to demonstrate beyond doubt that students come to physics teaching with preformed ideas about the issues at stake. They come in other words with an everyday physics. Second, although this everyday physics is sufficiently unorthodox to subvert the teaching process, it is not entirely lacking in contact with received science wisdom. In particular, it links with science orthodoxy over: (a) its reliance on variables, some of which are relevant; (b) its use of mechanisms, some of which are partial versions of received ones; and (c) its integration of mechanisms and variables, on some occasions at least, into theoretically organised structures. Third, this pattern of partial adequacy within an essentially theoretical structure is exactly what would be expected by those cognitive psychologists who wish to solve the problem of induction by calling upon the generative power of mechanisms. By virtue of this, the educational problems caused by everyday physics may also be more easily overcome than might initially be supposed. The consistency between everyday physics and the approach to induction is potentially significant, for we now know that endorsing the latter is an empirical issue. However, is the consistency compelling? This is the issue that remains to be seen.

2 Rationale for a developmental perspective

Chapter 1 covered two main issues. First, it demonstrated the existence of everyday physics in novice students of physics, and established something about its nature. Second, it showed how the nature of everyday physics is consistent with the claim made by certain cognitive psychologists and philosophers that theoretical structure is a primitive of human cognition. In particular, if theoretical structure is primitive, it will operate as an organising principle from early in life. Thus, it is consistent with the claim that the mechanisms and variables of everyday physics appear, to some extent at least, to be theoretically organised. In addition, if theoretical structure is primitive, there should be some internally generated pressure to understand mechanisms, since theories revolve around mechanisms. Thus, it is also compatible with the claim that the mechanisms of everyday physics are not entirely wayward when judged by received standards.

The claim that theoretical structure is primitive was advanced within cognitive psychology and philosophy as a solution to the classic problem of induction. This problem amounts to the fact that although there are an infinite number of ways in which reality can be segmented, we converge with minimal reflection upon some segmentations rather than others. As Chapter 1 explained, making theoretical structure primitive is a more acceptable response to the problem of induction than many obvious alternatives. Indeed, some of the alternatives call upon theory surreptitiously. Yet despite this, Chapter 1 did not succeed in proving the primitive status of theory. All the arguments that it was able to amass presupposed a perceptual basis to human cognition. Action has also been seen as central, and the implications of action for both theoretical structure and induction are currently uncertain. Because of this, the status of the findings regarding everyday physics that were presented in Chapter 1 is also unclear. Do these findings provide decisive evidence for the primitive status of theories and the mechanisms that these imply or do they not? A major aim of the present chapter is to answer this question.

To proceed, the chapter will start by exploring conceptual structure on the assumption of action-groundedness. As the discussion progresses, it will become clear that, on this assumption, theoretical structure cannot be seen as primitive. On the contrary, an action-based perspective demands that mechanisms be derived from thinking which is initially centred on variables. Thus, far from being generated by mechanisms as theoretical structure implies, variables are in fact prior and arguably foundational. Moreover, although action-groundedness allows for theoretical structure by the age at which physics teaching typically begins, it does not require this and it certainly does not anticipate it early in life. Furthermore, it predicts that for some time after theoretical structure is in place there will be variables which are not generated by mechanisms. Faced with such predictions, it will be obvious what answer must be given to the question posed above. No, the findings from everyday physics that have been presented so far do not provide decisive evidence for the primitive status of theories and the mechanisms they imply. These findings related to novice students of physics, and by the age at which physics teaching begins theoretical structure may have emerged without having been primitive.

In leading towards this conclusion, the discussion of action-groundedness will of course have signalled the way forward: research with younger students who are below the age of formal teaching in physics. Indeed, it will have emphasised the significance of such research by showing how it relates to competing conceptions of human cognition, one centred on action and the other on theory. However, does the research with younger students have to relate to everyday physics as opposed to other domains, and whether it does or not, to what extent has the research actually been conducted? The latter part of the chapter will revolve around these questions. It will be argued that although research with everyday physics is not essential, such research has a potential for clarity which would be hard to achieve with, say, everyday biology or everyday psychology. Certainly, the need for the research has not been pre-empted by work in these other domains, and thus the chapter will end by proposing a focus on young students' thinking about physics. By virtue of this, the chapter will hopefully have clarified and justified the claim made early in Chapter 1: everyday physics does indeed have psychological and educational significance but this is particularly the case when it is studied in the early years.

The chapter will end then by proposing that the work with novice physicists needs to be supplemented with research where children and younger adolescents are engaged in physics problem solving. The aim of the research would be to map age-related changes and/or continuities with regard to variables, mechanisms and, most importantly, variable–mechanism relations. Advocacy of this essentially developmental approach will set the scene for the next six chapters for all in some sense or another are attempts to present the current state of play as regards the research. The result is that the bulk of the book will be concerned with children and young adolescents, and thus may seem some steps removed from physics education as currently constituted. The point to remember is that the developmental approach is not being followed for its own sake, but rather as the optimal strategy for clarifying cognitive theory and advancing educational practice. Although the developmental approach is the means, cognitive theory and educational practice are most definitely the ends.

CONCEPTUALISATION AS AN ACTION-BASED PHENOMENON

The idea that cognition is grounded in action has found particular favour in continental Europe, with Great Britain and North America being more strongly influenced by the Humean tradition which emphasises perception. Yet, although the idea is popular throughout continental Europe, there are few attempts to spell out the implications for theoretical structure and/or causal mechanisms. Of the attempts which do exist, the most comprehensive is the one associated with Piaget, and thus it is with Piaget that the present section will begin. Piaget's account of action-based cognition will be outlined in detail, paying special attention to his claims about variables, mechanisms and their interrelation. Piaget's account will then be evaluated, though not at this point in the fashion that is familiar to psychologists. In particular, the issue will not be whether Piaget's account is supportable by evidence, but whether it defines the action-based approach or offers one option amongst several. The line taken will be that it does a little of both. Hence, the section will continue by considering other possibilities within an essentially Piagetian framework, looking especially at the tradition established by Vygotsky. Having moved by these means to a picture of what action-based cognition implies, the section will return to the major issue, the adequacy of everyday physics as outlined in the previous chapter to establish theoretical structure as primitive within human cognition.

Piaget and causal development

Piaget's views about cognition stem from his belief (detailed in Piaget, 1953, 1954) that at birth the child is endowed with a very small number of action patterns or, as he prefers, 'schemes'. Thus, development must involve the differentiation of these initial schemes into something more specific, and it was Piaget's understanding of how differentiation proceeds that proved to be crucial. As Piaget saw it, the schemes present at birth are activated by any entity (that is person or object) stimulating the relevant part of the body. Thus, any entity touching the mouth will activate the sucking scheme. Having been activated, the scheme will be applied automatically. To the extent that application is successful, the entity will be incorporated into the scheme, with important consequences. Since nipples, teats and fingers are more readily sucked than gloves, fists and blankets, segmentation of reality is already under way. However, the segmentation is dictated by the properties of action, and not by perception or theory. Moreover since the crucial actions are few in number and, as a biological necessity, of a particular kind, the form of segmentation is closely constrained. This in a sense is Piaget's solution to the problem of induction.

The segmentation imposed by action will not of course be static. To the extent that application of the activated scheme is unsuccessful (say, insufficient pressure is applied to hold the teat in place), an attempt will be made to modify the scheme to ensure success in the future. The process of modification which Piaget termed 'equilibration' is what guarantees the differentiation of the schemes of birth into more refined structures. However, while differentiation is stimulated by the entities that the child experiences, it continues simultaneously to impose associations upon them. Pencils and crayons are seen as similar because they are incorporated into the 'scribbling' scheme which emerges from the global grasping that is present at birth.

Of course, very few entities will be incorporated into one scheme alone. Pencils are, for example, likely to be incorporated into the 'chewing' scheme (a derivative of sucking) as well as the 'scribbling' (a derivative of grasping). This 'cross-tabulation' led for Piaget into the 'co-ordination' of schemes, a process deemed to have important consequences. In particular, by virtue of being associated with several schemes, entities develop autonomy from any one scheme, and thus take on an existence independent of action. This for Piaget was the central feature of development during the first two years, the growing awareness that entities which are known through action have an independent existence. This awareness was supposed to be manifested in the discovery of 'object permanence' a concept which Piaget famously denied at birth. It was likewise thought to underpin symbolic representation, as the child conjures up entities in the absence of overt actions. Combined with symbolic representation, it was believed conversely to give children the ability to work through actions mentally, in short to think. Finally and most significantly for us, it was supposed to provide insight into causal mechanism.

In his 1954 book, Piaget discussed in detail the child's conception of causality during the first two years. He pointed out that if at birth entities only exist by virtue of actions, entities cannot be deemed to initiate actions. Thus, the child cannot recognise him/herself as the instigator of actions, being limited to a feeling of 'efficacy' as actions occur. Conversely if actions are only conceptualised by virtue of their application to entities, causal relations between actions cannot be appreciated. All that is possible is a 'phenomenalistic' association between one action and another, between say engaging with the nipple and sucking the milk. As entities develop autonomy from actions, there should be a decline in both efficacy and phenomenalism, as the child becomes aware of self as mediator, and by virtue of this links actions as causes and effects. With further separation between actions and entities, the child can also begin to appreciate that other entities can be called upon to mediate effects, that by pointing to a toy the mother can be enlisted to help with its retrieval. Piaget dated this level of awareness at around the first birthday, and he anticipated subtle changes in it over the next few months. Initially, the causal power of external entities is seen as completely at the mercy of the child's activity. The child is a magician who conjures a helpmate. As the second year progresses, appreciation grows of the autonomous causality of others. Thus, entities apart from the child are seen as instigating actions in the child's support.

It should be clear already that Piaget completely reversed the line followed by Murphy and Medin (1985) and Harré and Madden (1975) as discussed in the previous chapter. For Piaget, the processes that dictate segmentation of reality, namely entity–action interactions, precipitate the processes which lead many months later to the discovery of causal mechanisms in the form of first own action and then action of another in support of self. As we saw in the previous chapter, the discovery of causal mechanisms is, for the other theorists, the primitive that precipitates the segmentation of reality. The difference is accentuated by the fact that Piaget saw developments after the second birthday as manifesting the same sequence of segmentation followed by mechanism as developments before. To appreciate why, note that everything discussed about the first two years relates to causality in the service of the child's activities. There is no suggestion that children under 2 can appreciate the causal effect of one entity on another in contexts where they have no vested interest. This is no accident, for Piaget not only denied awareness here, but claimed also that when 'the child can no longer structure reality by placing his own action among causes and effects arranged in a system external to it, he again confers on efficacy an unwanted power' (Piaget, 1954: 348).

The prediction is, then, that with events external to the child, the only mechanism recognised is the child's activity. As a result, the relation between events when the child's activity cannot be imputed must be purely phenomenalistic. Piaget saw the work that he conducted both before and after the 1954 book as providing ample confirmation that this is what happens. One example of the work appears in Piaget (1930). This book reports a series of studies where children aged 4 to 12 were interviewed about a range of physical phenomena, all external to themselves. These included the movement of clouds, the rippling of water, the floating of boats, the formation of shadows and the operation of bicycles. The children were asked to reflect on the phenomena, and explain why they happened. Up to the age of 6, the children's responses frequently contained elements of phenomenalism and personal efficacy, and even after 6 such elements were not unknown. A typical phenomenalistic response was to say that shadows are the joint product of the shadowed object and shadowy things like the night and dark. A typical efficacy response was to say that the child's walking causes the clouds to move. Similar evidence appears in Piaget (1974), despite lacking the rich descriptive element of the earlier work. For example, there is a discussion of 'transductive reasoning' in children up to 6, which turns out to mean juxtaposing events without recourse to mechanisms, or phenomenalism by another name. There are also examples of efficacy, as when children up to 6 call on the human voice to explain the transmission of sound.

The transcendence of efficacy and phenomenalism around the sixth birthday was believed by Piaget to involve processes which mirror those proposed for the first two years. Thus, as actions and entities become still further differentiated, the child who believes 'If I point, mummy will help me reach that toy' will co-ordinate that belief with 'If I scream, mummy will help me reach that toy' or perhaps 'If my sister points, mummy will help her too.' Likewise, the child who believes 'If I walk, the clouds will move' will co-ordinate that belief with 'If I run, the clouds will move' or 'If my sister walks, the clouds will also move'. In these circumstances which Piaget termed 'decentration', the consequent will become decoupled from any particular antecedent, giving it a degree of autonomy from the child's activity. With autonomy comes the possibility of recognising causal mechanisms which are completely independent, with a corresponding decrease in efficacy and phenomenalism. The point is that efficacy and phenomenalism were not believed to be transcended until around the age of 6. Thus up to the early years of primary school, Piaget is painting a picture that is diametrically opposed to our earlier theorists. Variables are acknowledged to be sure, for instance 'pointing helps me get things'. However, these variables are egocentric, and they cannot be constrained by mechanisms because mechanisms quite simply do not exist.

Beyond 6 years of age, the gap between Piaget and the previous theorists appears superficially to close. Mechanisms are seen as mediating within physical events, and these events are believed to be constrained by decentred variables. (Thus, we have 'Walking makes the clouds move' and not 'My walking makes the clouds move'!) However, while this convergence of perspectives must be accepted, mechanisms are not seen as cognitive imperatives within Piaget's theory, and thus children are not regarded as intrinsically motivated towards understanding. In addition, and more importantly, there are according to Piaget barriers towards bringing whatever is discovered about mechanisms to bear on variables. As Piaget pointed out, to ask 'Which variables are entailed by these mechanisms?', children must suspend belief in the variables they are currently endorsing. As Inhelder and Piaget (1958) put it, they must treat reality as an extension of possibility. Piaget realised that on his model this 'possibilistic' perspective could not be contemporaneous with the mere awareness of mechanisms for it implies another level of differentiation. In particular, it requires more differentiation between actions and entities to treat known action–entity relations as options than to take them for granted. Inhelder and Piaget located a possibilistic perspective no earlier than 11 or 12.

Piaget is implying, then, that prior to 11 or 12 cognition could not consist of neatly packaged theories. Certainly, it could refer to variables and from 6 onwards it could also refer to mechanisms. There may even be an association between variables and mechanisms, for children may refer to their beliefs about variables if they make an effort to understand how mechanisms work. (They may of course also look to sources apart from variables.) However, a situation where variables are generated by mechanisms would, for Piaget, be inconceivable. After 11 or 12, appreciation of the generative power of mechanisms is theoretically possible, with the consequence that children may now seek the variables that their mechanisms entail. However, the image of cognition as neatly packaged theories is unlikely to prove valid, for the question 'Which variables are entailed by my mechanisms?' does not necessarily imply the question 'Do my mechanisms entail the variables that I currently endorse?' As a result, the likely scenario after 11 or 12 is a composite system where some variables are theoretically grounded and some are outliers. This is not a system where science teachers could direct their energies at mechanisms in the knowledge that variables would take care of themselves. Equally, it is not a system where conceptual coherence is given by theory.

The epistemological status of Piaget's theory

Piaget tells a good story, even if it is not the story that science education would wish to hear. However, what is its significance? Is it unique to Piaget or does Piaget speak for all scholars who ground cognition in action? To answer the question, it would be gratifying to find other theorists tracing causal development on the assumption of action-groundedness. Unfortunately, such theorists do not exist. The Russian psychologist Lev Vygotsky probably comes closest in that he presumed action-groundedness and made extensive comments about cognitive development. His work is relevant and it will be discussed in due course. However, it cannot be said to address causality in detail nor to trace development from the first moments of life. Thus, there is nothing comparable in scope to Piaget which answers the same question, and this places constraints on how we proceed. Probably the best strategy is to begin with Piaget's beliefs about the first months of life, and ask whether his claims about actions are empirically well founded. Assuming they are, the next step will be to consider whether the remainder of the Piagetian edifice falls out logically from what can be presumed at the start.

Piaget's image of the very young child centres around a limited repertoire of actions that is completely bound up with the triggering entities. It is an image that was based on casual observations of his own three children, an inadequate database by any account. Recognising the inadequacies, there have been numerous attempts at supplementary research with, at first sight, varied results. The variation is however misleading for much of the research has utilised perceptual measures, visual tracking, duration of gaze and so on. Such research may reveal whether Piaget's emphasis on action led him to underestimate what children know. Nevertheless, it does not reveal what their significant knowledge is granted that everything of significance occurs through action. This is our present concern and to address it we need studies which have monitored action directly. Relevant studies exist, though interestingly they are amongst the earliest of the follow-ups to Piaget's research. Typical is the work of Gouin-Décarie (1965) and Uzgiris and Hunt (1975), which used tasks developed from Piaget, but administered them to larger samples in more controlled settings. The work offers impressive support to Piaget, and not just for the early months but across the whole of the first two years.

In supporting Piaget throughout the first two years, the work is in one sense doing more than we need. If we hold an action-based model of knowledge and if actions are limited in number and inseparable from entities, then it surely follows logically that the next stage of development should be as Piaget proposed. In which case, empirical evidence may be reassuring but it is not required. In particular, phenomenalism and the experience of efficacy will be all that is initially possible. Moreover self will be the first mediator to be acknowledged subsequently, with corresponding problems for external causality. Indeed, the sequence from 'helpmate under own control' through 'helpmate in support of self' to 'independent source of causation' can probably be hypothesised. What, though, about development after 2? Here too there seems a relentless logic to what Piaget proposed. Given the sequence before 2, it seems inevitable that the discovery that external entities can initiate actions in support of self must precede the discovery that external entities can initiate actions upon each other. Given this, it also seems to follow that mechanisms should cause more problems than variables, and that the generation of variables from mechanisms should offer particular challenges.

Nevertheless, while Piaget's logic is seductive, a number of questions must be raised. One relates to Piaget's tendency to *explain* development in the same terms as he *describes* it. In particular, development is described in terms of action differentiations and explained in terms of action co-ordinations (or, after 2, belief about action co-ordinations). While the latter may be defensible, it is not necessitated a priori by the former. Co-ordinations between actions and something apart from actions might in principle operate as the differentiating mechanism. Recognising this, it seems legitimate to ask whether there are other co-ordinations which children could make, and if there are whether they significantly affect the course of development. Researchers in the Vygotskyan tradition would answer 'Yes' to both questions, meaning that the moment has come to consider their work.

As mentioned previously, the Vygotskyan tradition is an action-grounded one, stemming in fact from a combination of Marxist philosophy and reports which Vygotsky read of research with chimps! If the point made above is correct and Piaget's claims are in some respects matters of logic, we should expect therefore to find echoes of Piaget within Vygotskyan writing. In the case of Vygotsky himself, his untimely death meant that he was unlikely to have been familiar with Piaget's claims about infant development. However, his biographer Kozulin (1990) leaves us in no doubt that he would have approved, writing that the observations on which Vygotsky's views were founded 'were brilliantly confirmed by Piaget' (Kozulin, 1990: 156) through his studies of the first two years. As regards the period from 2 to 6, Vygotsky did not, as we noted earlier, make detailed pronouncements about causal development. However, he did express views about general conceptual knowledge at this age level (see Vygotsky, 1962), these views being strongly influenced by empirical studies that he himself conducted.

Particularly significant were studies using a sorting task. This task involved twenty-two blocks varying in colour, shape, height and size that were scattered over a room. Each block had a nonsense syllable written on its underside, and one of these was displayed to reveal the 'name'. Children were asked to select the blocks that might have the same name, the nonsense syllables on the chosen blocks being subsequently displayed as feedback. Typically, there would be repeated cycles of selection followed by feedback. In his 1962 book, Vygotsky gave a detailed account of performance on the task. One key point is that the youngest children 'compensate for the paucity of well-apprehended objective relations by an overabundance of subjective connections and . . . mistake these subjective bonds for the real bonds between things' (Vygotsky, 1962: 60). From the quote it is clear that Vygotsky, like Piaget, saw children as starting from the viewpoint of self.[1] If as argued earlier, this viewpoint entails the Piagetian line on causal connection, we can safely assume that Vygotsky would have followed this too. Indeed, we can find a few lines in the 1962 book where Vygotsky came close to saying this. Discussing 'primitive thought', Vygotsky made one of his rare direct references to causality, and expressed his acceptance of Piagetian phenomenalism (citing Piaget explicitly). He wrote of the close interdependence recognised by children and (he hypothesised) primitive people 'between two objects or phenomena which actually have neither contiguity nor any other recognisable connection' (Vygotsky, 1962: 71).

So far so good, but what about the co-ordinations which drive development forward? As noted already, these are the main point of divergence between the Piagetian version of action-groundedness and the Vygotskyan, and the divergence is most marked over the processes by which children transcend their own point of view around the age of 6. For Vygotsky (see, also, Vygotsky, 1978), the processes were mediated by culturally shared symbols, which children come to impose on their action-based world. Imposition provides a level of detachment from the activity itself, which allows for its transcendence and regulation. Thus while Piaget was stressing co-ordination between beliefs about actions (about variables in this chapter's terms), Vygotsky emphasised co-ordinations between beliefs about actions and culturally shared symbols. The key consequence is that for Vygotsky, unlike Piaget, children are under pressure to submit their beliefs to cultural wisdom. This ensures a direction to development which is absent from Piaget.

Application to everyday physics

Taken as a whole then, Piaget and Vygotsky are very similar, and their points of similarity help us understand what is intrinsic to action-based theorising. However, Piaget and Vygotsky also differ, and the differences may mean that we are not dealing with a monolith when we address the implications of action-based theorising for everyday physics. To see whether we are, the best strategy is probably to take the main source of difference between the two theorists, namely culturally shared symbols, and see whether it implies divergence in the particular case of physics. Proceeding on this basis, the culturally shared symbols of relevance to everyday physics centre, surely, on language, in particular language reflecting commonplace beliefs about the physical world. This means, of course, language whose properties we can partially predict. From the discussion in the previous chapter of how everyday physics is constituted, we can anticipate language which sometimes mentions variables, sometimes mentions mechanisms and sometimes puts variables and mechanisms into a theoretical interrelation. Thus, it is this kind of language that we need to consider when asking whether Vygotsky makes predictions which differ from Piaget's.

To clarify the situation, consider, first, variables. From our earlier analysis of novice students of physics, we know that from the teenage years the variables which constitute everyday physics are partly but not entirely discordant with the received wisdom of science. This should be reflected in the language used by teenagers and adults when mentioning variables in conversation with children, implying, say, 'The wind's making those clouds move' and not 'Clouds move because people walk.' Since, as noted already, children may easily believe the latter around the age of 6, there will be conflict between the language experience and what is presumed, and the conflict will constitute pressure towards the variables which older individuals subscribe to. Given the relative (only relative) orthodoxy of these variables, this will amount to pressure to improve.

Within Piaget's theory, there does not appear to be equivalent pressure. Certainly, Inhelder and Piaget (1958) argue that with decentration children become capable of responding to feedback. Thus, they should notice that 'Walking makes the clouds move' is not always consistent with observed events. However there seems nothing in Piagetian theorising to impose immediate constraints on the variables which are selected instead. It is true that decentration around 6 leads, according to Piaget, to awareness of mechanisms. Thus, there might be an expectation of primitive variables being replaced by variables which concur with the operation of mechanisms. However, this could not happen on a Piagetian account. In the first place, 6-year-olds are not expected to know anything about the *nature* of mechanisms, simply their existence. In the second place, using mechanisms to guide decisions about variables implies a theoretical perspective which, as we have seen, Piaget denies until 11 or 12. The implication, if we return to the problem which has been guiding much of our discussion, is that, for Piaget, induction as regards variables is unrestricted between 6 and 11. It is only after 11 that Piagetian theory predicts the constraint and possible improvement which Vygotskyan theory allows from 6.

A similar difference emerges in relation to mechanisms. For reasons exactly parallel to those that have been argued for variables, children are likely to hear partial truths if they experience mechanisms represented in language. On the Vygotskyan model, this will help to push them in the right direction, a push which is not necessitated by Piaget's theory. However, in addition, linguistic representations of mechanisms may sometimes also signal a theoretical relation with variables. For instance, children could be told that it is to stop the heat getting out that the windows and doors are shut, and that it is to get the power up to strength that radios are loaded with batteries. To the extent that this happens, children should, on a Vygotskyan account, appreciate that mechanisms can generate variables much earlier than Piaget would anticipate. This is not because the question 'Which variables are entailed by my mechanisms?' is any easier on a Vygotskyan account than it is on a Piagetian. On the contrary, it is as argued already problematic for any theory which centralises action. Rather, it is because linguistic experiences protect children from having to ask the question in the first place.

This said, children are not protected from asking the second of the questions mentioned earlier in relation to Piaget, 'Do my mechanisms entail the variables that I currently endorse?' On the Vygotskyan account as on the Piagetian, children should have difficulty with this question until sometime after 6. The consequences should be twofold. First, even after theoretical awareness is in place, it will be difficult to bring all variables within the scope of mechanisms. Second, given outlying variables, it will be difficult to exclude notions that are scientifically irrelevant. As mentioned already, the Vygotskyan perspective permits the introduction of relevancies from direct experiences of language. It also permits this as an indirect consequence of theoretical awareness, for knowledge of the mechanisms which generate variables may also improve over time. However, until children look back over their beliefs as a whole, inclusion of relevancies will not necessarily be accompanied by exclusion of irrelevancies. Thus, for Vygotsky, as for Piaget, the latter will prove difficult.

Given the contrasts between Piaget and Vygotsky, it is clear that there is considerable scope for variation over everyday physics within the action-based framework. Nevertheless, the points on which the two theories concur seem likely to be true of any conceivable action-based model, namely the predating of mechanisms by variables and the partial autonomy of variables even after mechanisms come to be generative. These points of concurrence are however in complete contrast with the models introduced in the previous chapter, the models associated with Murphy and Medin (1985) and Harré and Madden (1975) and perhaps denotable as 'theory-based accounts' for relatively straightforward comparison with their 'action-based' counterparts. Theory-based accounts place mechanisms as prior to variables, and posit all variables as consequences of mechanisms.

Faced with contrast between theory- and action-based accounts (not to mention the variation within the latter), it will be even more obvious than it was in the previous chapter that mechanisms and theoretical structure cannot be *presumed* to be central to human cognition. Empirical evidence is most definitely needed, meaning that now is probably the moment to reconsider our earlier sample of research into everyday physics and ask about its decisiveness. Does it not merely mesh with theory-based accounts but also exclude action-based accounts? As signalled earlier, the answer is straightforwardly 'No'. First, to reiterate a point that has been made already, few students embark on physics before the teenage years. Thus insofar as our sample of research was concerned with novice students of physics, it was restricted to this age group or older. As a result, we are talking of a group where theoretically structured knowledge is predicted by theory-based accounts and allowed by action-based accounts. This renders evidence for theoretical structure somewhat ambiguous. Second, although the accounts make differing predictions even after theoretical structure has become entrenched, these differences cannot be tested with the earlier sample of studies. The differences relate to the existence of outlying variables, that is variables which are beyond the scope of mechanisms. To establish the true situation as regards such variables, it would be necessary to plot the relation between *every* variable and *every* mechanism that the students used. None of the studies sampled earlier attempted to do this. Indeed, it was noted in passing when the studies were discussed that only the work with electricity and heat transfer plotted the relation between *any* variables and the available mechanisms.

Is the solution then further, more focused research involving novice students of physics? I think not. Certainly if such students displayed variables which were not theoretically derived, it would be possible to endorse the action-based accounts over the theory-based ones. However, if they did not do this, would it really allow the endorsement of the theory-based accounts over the action-based ones? Could we ever be certain that we were not making our observations at too late a date, after the outlying variables that hitherto existed had finally been eliminated? Could we even be certain that outlying variables were not still in existence and being eclipsed by the demand characteristics of the study? Drawing conclusions from *absences* is a risky strategy, and only to be attempted if nothing else is possible. Clearly though something else is possible in the present context. As noted above, there is a key difference between the action- and theory-based models as regards development before the age at which physics is typically taught. There are also differences regarding development at younger age levels between the Piagetian and Vygotskyan versions of action-based growth. Suppose then that we were to work with a younger age group, and more particularly to trace development over time. Then, surely, we should be able to say something that is reasonably conclusive as regards both conceptual structure and its educational implications.

A STRATEGY FOR DEVELOPMENTAL RESEARCH

In view of the preceding section, there can be little doubt that a developmental approach could resolve some theoretically and practically significant issues. However, what form should the approach take, and can it be reliably implemented in practice? The basic point to note in relation to the first question is that the issues which have been raised for the preschool-age group are different from the issues raised for the school-age group. For preschoolers, the fundamental issues surround awareness of causal mechanisms with events where children are not personally involved. The theory-based models treat awareness as a cognitive primitive, and thus are committed to its existence from a very young age. The action-based perspective on cognition necessitates problems with causal mechanisms when the events are impersonal. Thus, awareness is impossible in the early stages, and may be delayed until as late as 6.

For school-age children, the fundamental issues relate to the integration of mechanisms with variables. The theory-based models predict that mechanisms will, at all ages, be seen as generative, and hence that beliefs about variables will always be contingent on knowledge of mechanisms. The action-based models deny this, regarding a generative perspective as problematic and incapable of binding variables until late in development. However, there are differences within the action-based camp. The Piagetian approach treats a generative perspective as impossible until around 11 or 12. Moreover, it is sceptical about whether understanding of variables and mechanisms taken separately will improve much before that age. The Vygotskyan approach is less rigid on both counts, permitting some grasp of generativity from 6 and allowing systematic improvement in variables and mechanisms. Confronted with the fundamental issues, the problem is how to research them, and addressing the problem will be the main aim of this section. The section will start with the preschool-age group, discussing the feasibility of studying awareness of causal mechanisms. Coming to a somewhat pessimistic answer, the section will turn to school-age children thereby setting the scene for the chapters to follow.

Research with preschool children

The psychological literature contains several studies with infants where events relevant to physics were presented and behaviour was observed which indicates sensitivity to causal relations. For instance, Leslie (1984) showed 6-month-olds a film where a red brick moved from left to right until it made contact with a stationary green brick, at which point the red brick stopped and the green brick moved to the right. This is of course equivalent to the procedures used by Michotte (1963) which, as noted in the previous chapter, give adults a strong sense of causal relationship. In Leslie's study, repeated presentations of the sequence ensured that the infants were familiar with it, whereupon new sequences were introduced which maintained some elements of the original one but altered others. One such sequence maintained the 'causal' structure but reversed the components, that is the green brick moved from right to left until it touched the red one. The infants showed more interest in this sequence than in others which, say, violated the causal structure while continuing the elements, suggesting that they were aware of the causality.[2]

By contrast, Baillargeon (1994) reports a series of studies with 3- to 10-month-olds where diminished interest was taken as evidence of causal knowledge. For example, in one study a hand reached behind a screen on two occasions, on each occasion to deposit a doll. The screen was then lowered to reveal two or three dolls. In one three-doll condition, the screen had been raised at the start; in the other, it was lying flat. The second three-doll condition generated considerable surprise and interest, while the reaction to the first three-doll condition was no different from the reaction to the two-doll. Baillargeon suggested causal explanation to account for the difference. Since the screen was flat at the start of the second condition, the doll could not have been there already, hence the amazement. It was, by contrast, perfectly possible for one doll to have been hidden behind the screen from the start of the first condition.

Such studies are fascinating, and there are several more of the same ilk. However, they are not strictly relevant in the present context, for they do not explore the *nature* of the causal link which the infants are presumed to be making. Is it mechanistic or is it phenomenalistic in the sense of Piaget? In other words, is the generative role of the cause appreciated, or is its linkage with the effect simply associative? These are the central questions for us, and it is not simply that infancy research has failed to address them. It is also that it is difficult to imagine how infancy research could proceed in order to address them. Certainly, as things stand, we have to turn to research with older preschoolers to find the questions raised explicitly.

Particularly influential amongst the research are the studies reported by Shultz (1982). Five studies were reported with the age range of participants varying in each. However, preschoolers figured prominently, and indeed one study included children who were as young as 2. The methodology in all five studies involved presenting simple effects relating to the transmission of sound, light and air, and asking which events from two possibilities were causing them. For example, the effect in one study was a spot of light, and the problem was which of two lamps was producing it. In a counterbalanced design, the lamps varied in spatial contiguity with the spot, in temporal contiguity, and in whether they were switched 'on' or 'off'. A lamp has to be switched 'on' to generate a spot of light. Thus, the fact that the 'on' lamp was preferred by children of all ages was taken as evidence for awareness of mechanism. Similar results and hence similar interpretations emerged from the other studies.

Shultz's studies are ingenious but they do not in my opinion establish awareness of causal mechanism. There is, surely, nothing to guarantee that the so-called mechanisms were different in epistemological status from variables. Thus, in the example given, there is nothing to preclude the possibility that the children were responding to the variable 'on-ness' vs. 'off-ness' just as they responded to the variables 'spatially contiguous' vs. 'spatially noncontiguous' and 'temporally contiguous' vs. 'temporally noncontiguous'. To establish causal mechanisms, Shultz would, I feel, have to investigate what the manipulations meant to the children, and this would almost certainly involve follow-up questioning. In other words, besides asking the children to indicate the cause, it would be necessary to ask them 'why' they responded as they did. The trouble is that follow-up questioning of this kind has been studiously avoided in work which, like Shultz, includes both preschoolers and events relevant to physics.

Suppose, though, that we relax the requirement that the events be relevant to physics. Psychological research with preschoolers has recently become focused on children's conceptions of biological and mental functioning, and in doing so has employed a methodology in which follow-up questioning has become the norm. Since the research talks freely of children holding 'theories', it sounds, despite its lack of physics content, extremely germane to our present concerns. Indeed if the case for theories is sound, research into everyday physics might be rendered redundant. After all, only theory-based approaches permit theoretical structure during the preschool years. Thus, evidence that such structure exists for biological and mental functioning would offer strong endorsement for theory-based approaches. Since by their very nature theory-based approaches cannot apply in some areas and not in others, the issues relating to physics would seem to be resolved.

As it turns out though, the research into biological and mental functioning does not make a conclusive case for theorising. With the biological research, this is because despite the follow-up questions we are no clearer about the subjective reality of the variable vs. mechanism distinction than we are from the work of Shultz. Take for instance research by Keil (summarised in Keil, 1992). In this research, preschoolers were typically presented with what from the professional perspective would be contrasting mechanisms, for example germs or poison as the causes of disease. They were then asked questions like 'Is X contagious?', 'Is it alive?', 'Does it think?' and 'If you chop it into tiny pieces, will it still make you sick?' Because the response patterns varied as a function of cause, sensitivity to mechanism was inferred. However, the response patterns could have varied if the causes were regarded by the children as variables and not in accordance with the professional perspective as mechanisms. Thus, the key issue for present purposes has been completely by-passed.

The research into mental functioning is, if anything, more problematic, for it is unusual to find causal mechanisms addressed let alone studied carefully. This neglect might be surprising because the work is typically subsumed under the 'theory' label, with reference in particular to 'theories of mind'. However it is traceable to a misleading 'foundational' declaration by Premack and Woodruff (1978) to the effect that 'In saying that an individual has a theory of mind, we mean that the individual imputes mental states to himself and to others. . . . A system of inferences of this kind is properly viewed as a theory, first, because such states are not directly observable, and second, because the system can be used to make predictions' (Premack and Woodruff, 1978: 515). The claims about theory are, of course, false in every respect, for countless knowledge structures apart from theories refer to nonobservables and can be used in prediction. Schank and Abelson's (1977) scripts and Holland *et al.*'s (1987) production rules are two examples that we have already discussed. (For further examples and a similar point, see Hobson, 1991). Because Premack and Woodruff's declaration has, despite its problematic nature, been the starting place for much of the recent research, there has been a focus on what children know about mental states and not what they believe about causal mechanisms.

This said, the mismatch between the 'theory of mind' label and the data associated with it has not passed unnoticed, with two types of reaction. One, epitomised by Whiten and Perner (1991), has been to replace the label, using something more neutral like 'mind-reading'. The other has been to scrutinise the data post hoc and consider whether there is a causal-explanatory flavour that warrants the 'theory' label in its proper sense. It is this second reaction that is of interest here; yet if we probe it more deeply we soon discover differences of opinion over what the data mean. At one extreme is Wellman (1988, 1990) who claimed that 'from a young age children share much of our causal-explanatory framework for human action' (Wellman, 1988: 79).

Wellman's evidence is a series of studies, by himself and others, into children's appreciation of how beliefs affect behaviour. Some studies involved scenarios where the protagonist's beliefs may have been correct, for example 'Sam wants to find his puppy. His puppy might be hiding in the garage or under the porch. But Sam thinks his puppy is under the porch.' Children were asked where Sam would look. Others involved scenarios where the protagonist's beliefs were manifestly false, for example 'Maxi watches as a chocolate is hidden in the kitchen. While Maxi is away, unbeknown to him, the chocolate is moved to the living room.' Children find the latter, false belief, scenarios relatively hard. Nevertheless, even here, they predict actions from beliefs well before the end of the preschool years. Moreover, when asked to justify their predictions, they invoke beliefs, for instance 'Maxi will look in the kitchen because that's where he thinks the chocolate's hidden.' It is this association of prediction, belief and the word 'because' that persuaded Wellman of the reality of mechanisms.

Working within the theory of mind tradition, Harris (1991) has challenged

Wellman's interpretation. He argues that children could perform as they do with the belief scenarios by projecting onto the protagonists the actions that they themselves would carry out given those beliefs. Harris talks of 'simulations' rather than 'scripts', but it is this kind of knowledge that he has in mind. He is undoubtedly correct to propose simulations as a possible model of children's activity. Without doubt, they fit the data as adequately as do Wellman's theories. Nevertheless, in focusing on children, Harris is missing a fundamental point, which is that 'X did Y because of belief Z' does not guarantee causal mechanisms even in adults. It would only do this if adults attributed causal powers to beliefs themselves, but most adults would reject this proposition as superstitious nonsense! It is true that if asked 'How did belief Z lead X to do Y?' adults would typically resort to mechanisms, something psychological like 'willpower' or more likely something crudely physiological. However, their recognition of mechanisms is not revealed in the belief statements per se.

Even researchers who follow Wellman have not, by and large, seen the mechanisms which translate mental states into physical action as deserving of study. There is however one exception: a paper by Johnson and Wellman (1982). It is odd to see Wellman as one of the authors, for the tenor of the paper is the great difficulties children have with the mediating role of mind and brain. The point is made that in the age range 5 to 11, young children focus on the mind as a repository of mental states and psychological characteristics. It is only older children who posit a controlling function for the mind and brain. Thus, there is a strong suggestion that causal mechanisms are not appreciated in the realm of human action until long after Wellman himself posited 'theories' of mind. Could the suggestion be correct? Certainly, other literature hints of something similar. Reporting a series of studies concerned with children's conceptions of bodily functions, Carey (1985a) commented that it is not until 10 years that children construe the body as like a machine. Likewise, Broughton (1978) found few children younger than 8 mentioning the mind or body as directors of self. Finally, Connelly (1993) offered 5-, 8- and 11-year-olds a forced choice between physiological mechanisms and psychological states as the causes of a 'learning difficulties' child's problems at school. There was a marked increase with age in the choice of physiological mechanisms.

These studies cannot be said to be conclusive. In the first place, there are not enough of them. In the second, their methodology does not cover all the options. Connelly, for example, did not include psychological mechanisms or physiological states. Nevertheless, the message of the studies is one of caution: there is certainly no evidence for causal mechanisms in children's conceptions of mental functioning and there is a little evidence against. Overall then, just as with the biological material, research into children's conceptions of mental functioning has not established causal mechanisms in the thinking of preschoolers and as a consequence does not bear incisively on the issue at stake. Does that mean, therefore, that more research with preschoolers is required? For reasons given in the next few paragraphs, it seems to me that the answer is 'No'.

Rationale for a school-age focus

There are, I think, two problems with taking a preschool focus, one methodological and one theoretical. Since both problems are reduced if we shift to a school-age population, this is what I should like to propose. On the methodological front, the main difficulty is the reliance on language. As argued above, follow-up questioning is essential to establish what children understand about the events they witness. Without questioning and hence without heavy use of language, the ambiguities of, say, Shultz (1982) would appear inevitable. However, the language dimension creates problems of its own, for regardless of whether cognition is grounded in perception or action, it is not based on language. Thus, by requiring children to display their knowledge in language, a level of abstraction is being introduced which seems likely to lead to underestimation. In particular, there is a danger of false negatives, that is the failure to display causal mechanisms when they are known. This would of course favour an action-based gloss being placed on the data. Shifting to an older age group would not eliminate the problems entirely, but it would undoubtedly diminish them. In the first place, the children's language skills would be greater. In the second (and more importantly), the questions that we have earmarked for school-age children are not focused on presence vs. absence. Rather, they are focused on how two aspects of knowledge develop when they are present and how they are interwoven. Thus, should one or the other aspect not be displayed, our sole option would be to reserve our judgment.

On the theoretical front, the main difficulty with preschool research is that it could, at most, only bear on the theory vs. action dimension. It could not relate to the differences within the action framework, to the differences between, say, Piaget and Vygotsky, for the predictions do not diverge here with the preschool-age group. As intimated already, the claims made by Piaget and Vygotsky are not distinctive until the school-age level. At the school-age level however, there is divergence not simply between Piaget and Vygotsky, but also between these individuals and their theory-based counterparts. This divergence is expressed in terms of a set of predictions within Table 2.1. Noting the methodological and theoretical difficulties with the preschool-age group (and of course the ambiguities with the late- and postschool), a strong case can, I feel, be made for focusing on the school-age situation as represented in Table 2.1. The proposal is, then, to adopt this focus from now onwards.

Accepting that the school-age population should become the focus, research is needed which allows us: (a) to chart beliefs about variables and mechanisms in children whose ages range from 5 to 6 up to the early teenage years, and to map any age-related changes; and (b) to assess the extent to

Table 2.1   Key predictions relating to school-aged children given the theory- and action-based approaches

	Theory-based approaches	Action-based approaches (Piaget)	Action-based approaches (Vygotsky)
Positions regarding variables	Understanding of variables will improve throughout the school years.	Improvement in understanding of variables will be modest until about 11 or 12.	Understanding of variables will improve throughout the school years.
Positions regarding mechanisms	Understanding of mechanisms will improve throughout the school years.	Improvement in understanding of mechanisms will be modest until about 11 or 12.	Understanding of mechanisms will improve throughout the school years.
Positions regarding variable–mechanism relations	Mechanisms will generate all variables throughout the school years.	Mechanisms will not generate any variables until about 11 or 12. Thereafter there may be some generation, but bringing variables within the scope of mechanisms will be gradual.	Mechanisms will generate variables throughout the school years, but some variables will lie out with the scope of mechanisms until about 11 or 12. Bringing variables within the scope of mechanisms after 12 will be gradual.

which beliefs about mechanisms dictate choice of variables during the age range of interest. In theory, the research does not have to relate exclusively to physics, for just as with the preschool issues, evidence relating to Table 2.1 could come from the biological and/or the psychological domains. Indeed, if the evidence from these domains was compelling it would pre-empt the need for research with physics. However, the fact that conclusive evidence from one domain would be sufficient to resolve the issues seems to recommend an in-depth analysis of one literature rather than a broad-brush approach to several. Moreover, given that focusing is desirable, a case can be made for preferring to work with physics. At the very least, the advanced nature of the professional science means that improvements in understanding will be easier to detect.

The proposal is, then, not simply to address the issues identified in Table 2.1, but to do this within the context of everyday physics. This is in fact the agenda for the remainder of the book, the idea being to locate relevant studies and see what they have to say about the issues at stake. As it happens though, even within physics, there are too many studies to be considered in the space of one book, meaning that choice also needs to be exercised within the domain. This is where the studies introduced at the start of Chapter 1 once more become relevant, that is the studies concerned with novice students of physics, for these studies provide hints as to the basis on which choice should be made.

As noted already, evidence indicating theoretical structure in the early stages of physics instruction could only be obtained with a subset of the studies' topic areas. This could reflect nothing more than failure to ask the appropriate research questions. However, it could also reflect differences between topic areas over the extent of theoretical structure within the 'mature' knowledge. In which case, the theory-based approaches would be wrong, and this would be revealed in developmental profiles which departed from the left-hand column of Table 2.1: in some cases at least, mechanisms would not generate variables throughout the school years. What would remain to be seen is whether mechanisms would generate some variables throughout the school years or whether generation would wait in all cases until 11 or 12. Depending on the outcome, the pointers would be towards a Vygotskyan as opposed to a Piagetian version of the action-based approach.

It was a long shot but, in the absence of other guiding principles, I decided that the most revealing selection of topic areas might be one which referred to the evidence for theoretical structure at the time physics teaching begins. Reviewing the literature with this in mind, I found topic areas where theoretically structured knowledge is indisputable at the relevant age level, topic areas where it has been proposed but not without controversy, and topic areas where no claims have been made and no data presented. I decided to focus on one topic area within each group, giving me the three themes which are distributed across the next six chapters.

Heat transfer was chosen to represent topic areas which novice physicists treat in a theoretically structured fashion. It is a well-researched area with children aged 5 and upwards, and work like the Clough and Driver (1985a) study outlined in the previous chapter shows that for teenagers variables are indeed generated by mechanisms. Heat transfer will be discussed in Chapters 3 and 4. Propelled motion was selected to reflect topic areas where theoretical structure amongst novice physicists has occasioned debate. As we saw in the previous chapter, Gunstone and Watts (1985) and McCloskey (1983a, b) have used the theory analogy to interpret students' thinking regarding motion. However, their claims go, in reality, beyond their data. As intimated already, it was heat transfer and electricity which, amongst the sampled areas, provided evidence for theorising; propelled motion was sampled but recognised as ambiguous as regards theoretical structure. Noting the ambiguity, diSessa (1988, 1993) has strongly challenged McCloskey's views preferring to represent knowledge of propelled motion as fragmentary in character, comprised of 'phenomenological primitives' and not theoretical constructs. The debate here will provide a backcloth to the analysis of propelled motion that is the theme of Chapters 5 and 6.

Finally, object flotation is the choice from topic areas whose status amongst novice physicists is currently mysterious. As mentioned earlier, Piaget (1930) included flotation amongst his battery of topics, and this has stimulated a wealth of research with children up to 12. However, very little has been done with the slightly older age group, and hence we know virtually nothing about the ideas with which novices come to physics teaching. Clarification of thinking about object flotation will be the focus of Chapters 7 and 8. The guiding question across all of the chapters will be what children's thinking reveals about the predictions in Table 2.1, and hence, if the preceding arguments are correct, what is the message of everyday physics for cognitive theory and classroom practice.

Part II Heat transfer

3 Temperature change and childhood theorising

The two chapters which comprised part I of this book advanced arguments for adopting a developmental perspective towards everyday physics. In particular, these chapters tried to show that a developmental perspective should generate material of great relevance to both psychological theory and educational practice. However, while the appropriateness of developmental research was strongly defended, the chapters in part I did not advocate investigation across the full age range. They argued that the minimally verbal techniques favoured by some developmental psychologists would probably not permit unambiguous answers to the significant problems. Follow-up questioning would almost certainly also be required. However, the use of questioning with the preschool-age group could generate difficulties. At this age level, command of language is not necessarily secure. Moreover, the key issue with regard to preschoolers is of the presence vs. absence type, meaning that the danger of distortion due to language problems is particularly acute. With school-age children, language skills are obviously more advanced, and the key issues at this age level rest on language being used in one way rather than another. Acknowledging this, part I advocated research with school-age samples.

In detail, the key issues with regard to the school-age population are: (a) how knowledge of variables and mechanisms changes (or fails to change) with age; and (b) how closely the variable–mechanism relation resembles a theoretical structure. Part I introduced a set of cognitive models which it termed 'theory-based approaches' and which it saw as taking a straightforward stance on both of the issues. In detail, theory-based approaches predict that understanding of variables and mechanisms will improve steadily over the school years, within the limits of 'mature' everyday physics. This is because children see variables and mechanisms as theoretically related from a very early age, leading them to seek understanding of mechanisms and using this understanding to generate variables.

As part I pointed out, the stance taken by theory-based approaches has welcome implications for science education, but they are not for all that uncontroversial. A second set of models was also introduced in part I, these being referred to as 'action-based approaches'. Action-based approaches deny theoretical structure as a cognitive primitive, but rather see it as emerging (if it appears at all) from knowledge structures which rely initially on variables. As a result, variables will lie outside the scope of mechanisms long after theoretical awareness is established, and in fact well into the teenage years. This said, there is scope for difference within the action-based framework and, as part I made clear, this scope is well illustrated by Piaget and Vygotsky. On Piaget's version of action-based growth, understanding of variables and mechanisms is unlikely to improve significantly until 11 or 12, and theoretically structured knowledge is impossible before that age. On Vygotsky's version, both improvement and theoretical structure are possible from the earliest years of schooling.

In reality, Vygotsky's model does not simply permit improvement and theoretical structure from the earliest years of schooling; it also to a certain extent predicts these. For Vygotsky, development after 6 is driven by cultural-symbolic practices, which in the context of interest mean references to physics in language. As made by adults and older pupils, these references will reflect everyday physics as described at the start of part I. In which case, the references will be to variables and mechanisms which, despite inadequacies, possess elements of truth. These will be forces for improvement on the Vygotskyan model because they will far surpass the recently decentred notions that are anticipated at 6. Equally, the references will in some cases depict variables and mechanisms as theoretically related, and as such will press children towards theoretical structure. This said, there is no necessity from the evidence presented in part I that all linguistic references will manifest theory. It is even possible from the evidence that topic areas will differ here. If this is what happens in practice (and it is impossible at present to be certain), the Vygotskyan model would predict variation across topic areas in both the course and outcome of development. Such a prediction would bring the model even more firmly into conflict with the theory-based approaches, while maintaining the distance from Piaget. Noting all this, there is a clear need for research which not only covers a wide age range but also and equally importantly includes a broad spectrum of topics. Part I acknowledged both points by proposing an age range running from 5 to early teenage and identifying three distinctive topic areas.

One of the topic areas relates to the transfer of heat. It was chosen because there is strong evidence of theoretical structure in novice students of school and college physics. This is to say that students definitely appreciate that certain variables make a difference to how quickly substances heat up or cool down; they subscribe to mechanisms of heat (or cold) transfer which explain what is going on; and their espoused mechanisms are a major constraint on their choice of variables. Assuming that 'mature' knowledge takes this form, the developmental question is how does it emerge. The theory-based approaches predict steady improvement in variables and mechanisms and a constantly theoretical relation between these; Piaget predicts negligible improvement and no theoretical structure until after 11; Vygotsky predicts steady improvement and, given the nature of mature knowledge here, theoretical structure (although for Vygotsky variables beyond the scope of mechanisms would also be expected). Who, if anybody, is supported by the heat transfer data? This is the question that the present chapter and its successor will attempt to address. The present chapter will start by considering when children make the fundamental distinction between heating up and cooling down, that is when they acknowledge temperature change. Then the chapter will chart the variables that children deem relevant to temperature change, the mechanisms that they refer to, and most importantly the relations that they construe between variables and mechanisms. The next chapter will focus on the somewhat different case of changes of phase, that is the situations where solids turn into liquids, liquids turn into gases and vice versa. It is a celebrated principle of thermodynamics that although heat energy flows in such situations, there is no temperature change. The issue that the next chapter will address is whether children characteristically grasp this principle, and whether they do or do not, how their conceptualisations of changes of phase relate to what the present chapter will show about changes of temperature.

VARIABLES RELEVANT TO TEMPERATURE CHANGE

It is possible to think of a wide range of situations where substances are exposed to a heat source or removed from one, and a temperature change occurs. The situations can involve solids as with the pie in the oven or the wine bottle in the ice bucket, liquids as with the oil in the frying pan or the water in the hot water bottle and gases as with the helium in the hot air balloon or the flame in the Bunsen burner. In all these situations, energy is transferred from the hotter element to the colder, thus from the oven and the wine bottle in the first two examples. Energy will continue to be transferred until the two elements are at the same temperature, that is in 'thermal equilibrium'. Basically, there are three mechanisms of energy transfer: conduction, convection and radiation. Thus, when we consider children's mechanism knowledge and its relevance to variables, we shall be taking these three notions as the yardstick. However, to start, let us focus on variables alone and try to establish what beliefs are held about them during the age range of interest.

Temperature change as a conceptual distinction

The everyday lives of children are permeated with experiences of sources of heat. In summer, they feel the warmth of the sun's rays and on visits to science museums hear mind-boggling accounts of the heat at its core. At home, they encounter the heat of ovens, microwaves and barbecues in the context of cooking, and they feel cosy thanks to fires, radiators and underfloor pipes. As experienced in everyday life, sources of heat are both comforting and threatening. It is pleasant to lie out in the sun, but there is always a danger of burning. There is nothing more tasty than scones fresh from the oven, but if you eat them too soon they may hurt your mouth. As for fires, everyone enjoys being near them on a cold winter's night, but you must be on your guard for the sparks that fly out.

The threatening aspect of heat is almost certainly what informs the dialogues between young children and their parents. Studies of language development have revealed that words like 'hot' and 'cold' enter children's vocabularies at a very early age. However, suggesting extrapolation from contexts of threat, the words are often used for general prohibitions, without necessitating the presence of heat. For example, in my own work on early language (Howe, 1981), I remember a 20-month-old girl saying 'Hot, hot, hot' in relation to my video recorder. Since the recorder was at room temperature (and in any event she never touched it) she must have been referring to its forbidden nature. Other researchers have made similar observations, suggesting that a physical conceptualisation of hot and cold is an emergent phenomenon.

Once hot and cold are firmly grounded as physical properties, it becomes appropriate to ask how children think objects manifest one or the other. Are hotness and coldness intrinsic or are they acquired? In other words, are hotness and coldness qualities that objects do or do not possess, or can they be adopted by, for example, contact with hot and cold sources? Moreover, if hotness or coldness are acquired, is their acquisition instantaneous or does it take place over time? When children say the latter, it is possible to credit them with an appreciation of heating up and cooling down and by virtue of this to quiz them about the issues of central concern to this chapter.

Recognising the above, it is of relevance to consider work reported by Albert (1978). This work involved interviews with forty children aged 4 to 9. One question was 'What is the hottest thing in the world?' It is interesting that all the children treated the question as reasonable, indicating that hotness was construed by them as a physical property. The answers confirmed this, for the nominated objects were both physical and in fact hot. The sun (though hardly part of the world!) was a favoured response. Other questions related to the acquisition of hotness, with a stagelike progression being observed. The youngest children saw objects as intrinsically hot or cold, but by 5 or 6 appreciated that objects could become hot by, as it were, association. However, at 5 or 6, the creation or destruction of hotness appears to have been treated as instantaneous, for it was not until 7 or 8 that the children acknowledged becoming hot as extended over time.

Variables relevant to rate of change

It is then at around 7 or 8 that children see hotness and coldness as: (a) physical; (b) acquirable; and (c) gradual. Thus, it is at around that age that they can be said to appreciate the conceptual distinction between heating up and cooling down, and hence be questioned about the issues that the distinction entails. The first issue of concern to us is children's beliefs about the variables relevant to rate of change, and it is of interest that most published work addresses children considerably older than 7 or 8. There is for instance the work of Clough and Driver (1985a) which played such a significant role in part I. It will be remembered that Clough and Driver interviewed eighty-four 12- to 16-year-old pupils about the reasons for becoming hot and becoming cold. Their questioning centred on three issues: why metal spoons immersed in hot water feel hot, why metal plates feel colder than plastic ones, and why on a chilly day the metal part of a bicycle handle feels colder than the grip. Clough and Driver found that the presence or absence of metal was a key variable for many pupils: metal spoons let the heat in and metal plates and handlebars do the same for the cold. Two years later, those of the original sample who were still available were re-interviewed. As reported by Clough *et al.* (1987), the responses obtained in the second interviews were remarkably similar to those obtained in the first.

A picture is emerging, then, of an emphasis on metalness as the crucial variable and this would certainly square with other research. The findings of Clough and her colleagues are consistent with the survey reported by Brook *et al.* (1984) that was also mentioned in part I. As was explained in part I, the survey involved 900 pupils aged 11 to 15. On a smaller scale, though including for the first time the full age range of interest, Erickson (1979) interviewed a group of 6- to 13-year-olds about the results of, for example, placing objects on a hot plate and heating rods of different material. Erickson quotes extensively from what the children said, and references to metalness feature prominently. However, while the importance of metalness must be recognised, it is not the only variable to be cited in the literature. Approximately one-fifth of the responses to Clough and Driver's plates and handlebars items focused on 'appearance', in particular colour, thickness and smoothness. Moreover, Erickson notes how his respondents mentioned size, softness and strength.

Clearly then, children's beliefs in the domain of heat transfer are not univariate, but how much emphasis is given to each of the variables? Moreover, do children typically subscribe to one variable only or do they refer to a range? Because the literature does not even hint at answers to questions such as these, I attempted myself to obtain relevant data, through an interview study with pupils aged 6 to 15. Some 126 pupils were interviewed, with approximately equal numbers in the 6 to 7 age group, the 8 to 9, the 10 to 11, the 12 to 13, and the 14 to 15. The interviews covered a range of topic areas in addition to heat transfer, and thus will also be relevant to subsequent chapters. Noting this and noting also that the study has hitherto only been published in summary form (Howe, 1991), I have used the appendix to present full procedural details. As will be apparent from the appendix, the study deployed sixteen photographed scenes, with each scene being associated with a string of questions.

Two of the scenes are relevant in the present context. The first involved four pans sitting on a cooker, with one pan containing water. Of the questions associated with it, the following were intended to elicit beliefs about variables: 'Does the kind of pan make a difference to how quickly the water will heat up? Has the cook chosen the best pan for heating the water quickly? Which pan would be best? Why? Would the other pans be equally bad or would some be better than others? Why?' The second scene involved four forks around a lighted barbecue, with one fork being used to cook a sausage. Relevant questioning here was along the lines of 'Why is the cook holding a sausage with a fork? Do you think her fingers could still burn even though she's got a fork? Does the kind of fork make a difference to whether fingers will burn? Has the cook chosen the best fork to keep her fingers from burning? Why? Would the other forks all be as bad as each other, or would some be better than others? Why?'

In contrast to other research, my interview questions provided opportunities to deny that heating and cooling are influenced by the objects involved. As it happened, forty-three pupils did deny object relevance with the pans scene and fifteen did this with the forks. Denials were heavily concentrated in the two youngest age groups (χ^2 for pans = 24.52, df = 4, p<0.001; χ^2 for forks = 14.53, df = 4, p<0.01). Despite this, the majority of pupils at all age levels accepted that the involved objects were relevant, and were able to identify a best and worst set-up for the problem at hand. When they did this, they invariably selected one or more variables to justify their responses. Consistent with the literature, these variables included metalness, colour, thickness (or 'length' for the forks), smoothness, size, softness and strength. However, they also included very much more, for the pans scene elicited twenty-seven different facilitators of heating and the forks scene twenty-one different inhibitors of burning.

Yet within the heterogeneity, two variables stood apart, for both the pans scene and the forks. This was because they were used with very high frequency, being mentioned by at least thirty pupils when no other variable was mentioned by more than thirteen. With the pans scene, the high frequency variables were thinness and metalness as facilitators of heating. With the forks scene, they were length and non-metalness as inhibitors of burning. The striking thing about these variables is that they are scientifically relevant, for both thinness and metalness facilitate conduction from the heat source through the object. As a result, relevant variables predominated in the pupils' responses. With the pans scene, 65 per cent of the total references to variables called upon something relevant. With the forks scene, the corresponding figure was 84 per cent.

However, popular though relevant factors may have been, they were not equally distributed across the age groups. As Table 3.1 shows, there was a trend for the number of relevant factors to increase with age.

When the trend was analysed via ANOVAs, the results were in both cases statistically significant (F for pans = 11.88, df = 4,121, p<0.001; F for forks

Table 3.1   Mean number of relevant and irrelevant variables as a function of age: temperature change (Howe, 1991)

	6–7- *year-olds*	8–9- *year-olds*	10–11- *year-olds*	12–13- *year-olds*	14–15- *year-olds*
Pans					
Relevant	0.04	0.28	0.75	0.83	0.59
Irrelevant	0.54	0.40	0.33	0.29	0.52
Forks					
Relevant	0.54	0.80	1.21	1.13	1.48
Irrelevant	0.39	0.12	0.17	0.25	0.26

$= 9.73$, df $= 4,121$, $p<0.001$). Follow-up Scheffé tests indicated that the sample divided statistically into the two youngest groups versus the three oldest. The number of relevant variables was negatively correlated with the number of irrelevant (r for pans $= -0.21$, df $= 124$, $p<0.05$; r for forks $= -0.34$, df $= 124$, $p<0.001$). Thus, an age-related decrease might have been anticipated for irrelevant variables. However, as Table 3.1 indicates, this was not the case. Confirming this, ANOVAs on irrelevant variables as a function of age yielded no significant effects.

In revealing age trends over relevant variables, my 1991 study sets itself apart from the background literature, even indeed from Erickson (1979) which used a similar age group. Although age trends were not explicitly checked in previous studies, there are no indications in either the data or the reports that they might have been occurring. Also setting my study apart is the suggestion that thickness/length is a salient factor for children for, as explained already, it is metalness that is emphasised in earlier research. Can my results be accepted? Obviously, responses to a two-item 'test' must be treated with circumspection, but in the present context I have another dataset to bring to bear and the message here is confirming.

My second dataset derives from an investigation reported by Howe *et al.* (1995a). This investigation involved interviews with one hundred pupils aged 8 to 12, with exactly twenty-five pupils in each of the four relevant age bands. At the start of the interviews the pupils were shown eight contrasting containers, referred to by the interviewer as the thin cup, the black cup, the large bowl, the thick cup, the small bowl, the plastic beaker, the aluminium tin and the white cup. The containers were presented one-by-one and the pupils were invited to imagine that they had put given volumes of hot water into four of them and equivalent volumes of cold water into the others. They were then asked to predict whether the water would cool down (or heat up) quickly or slowly. Predictions were to be justified and justifications to be probed until the pupils' views had been fully explored. Thereafter, the pupils were invited to reflect on four real-world instances, for example a hot water bottle after a night in bed and hot chocolate after being left untouched.

Once more predictions were to be made regarding the rate of heating or cooling, with justifications elicited and probed.

The methodology differs from my 1991 study in several respects. More items were used, some items involved real objects and there was a greater emphasis on concrete prediction. In addition though, there was by virtue of both the appearance of the containers and the labels they were given deliberate directing of attention to four dimensions. In the Howe *et al.* paper, these dimensions are referred to as thickness, material (plastic or metal), surface area and colour (black or white). All four dimensions are scientifically relevant, the first two for reasons already given and the second two for reasons that are easy to grasp. In particular, with the volume of water held constant, the time to reach thermal equilibrium will be less with a large surface area than with a small one. It will also be less with a black container than with a white, in that dark surfaces absorb heat more rapidly than do light.

Because relevant dimensions were so explicitly signalled, coding for irrelevant variables could be misleading. As a result, the analysis focused on relevancies, attempting however to provide a comprehensive picture. Essentially, the responses to each item were coded at four levels along each of the dimensions. Failure to mention a dimension resulted in a score of zero; mention in the wrong direction (for example, 'Thickness helps water cool down quickly') a score of one; mention in the correct direction a score of two; and mention in the correct direction and co-ordination with another correctly used dimension (for example 'The metal will help the cooling but the thickness will make it harder') a score of three. Based on these scores, every pupil was awarded a mean score across items for each of the dimensions.

For present purposes, the interest is in how the mean scores varied with age and Table 3.2 provides the relevant data. Looking at the rightmost column of Table 3.2 it is clear that over the four dimensions scores did improve with age, an improvement which turned out to be statistically significant ($F = 7.91$, $df = 3,96$, $p<0.001$). Scheffé tests revealed that the 8- to 9-year-olds and the 9- to 10-year-olds had obtained mean scores that were significantly lower than the 11- to 12-year-olds with no other comparisons proving statistically significant. The results with thickness and material mirrored the overall picture closely with significant age effects (F for thickness = 4.64, $df = 3,96$, $p<0.01$; F for material = 6.39, $df = 3,96$, $p<0.001$), and a similar outcome after age group by age group comparison. (The 10- to 11-year-olds were however significantly different from the 11- to 12-year-olds with material.) With surface area, there was a trend in the direction of improvement with age but this did not prove to be statistically significant. With colour, we seem to have identified a dimension that virtually no pupils consider. Needless to say there was no age effect.

Relating the results back to my 1991 study, there are obvious points of contact. There too, there was an increase with age in references to relevant

Table 3.2   Four relevant variables as a function of age: temperature change
(Howe *et al.* 1995a)

	Thickness	Material	Surface area	Colour	Average across dimensions
8–9-year-olds	0.16	0.35	0.22	0.02	0.19
9–10-year-olds	0.16	0.68	0.29	0.05	0.29
10–11-year-olds	0.33	0.66	0.28	0.003	0.32
11–12-year-olds	0.56	0.84	0.47	0.02	0.47

variables. Moreover, it was an increase that took place to roughly the same time scale. In the 1991 study, there was a big leap in references to relevancies after a long period of no change between 8 to 9 years and 10 to 11. Here references to relevancies were rare at 8 to 9 years and 9 to 10 but much improved at 11 to 12. The significance of both trends for the action-based approaches of Piaget and Vygotsky will not have escaped notice, and will be discussed further later. This recognised, the picture across the two studies is not identical, for unlike the 1991 work the 10- to 11-year-olds in the Howe *et al.* study were not significantly different from the younger groups. The discrepancy could reflect differences due to demography for the children who featured in the 1991 study came from one of the most affluent areas in Scotland while the present children lived in a deprived, ex-shipbuilding area of Glasgow. Alternatively (or in addition) it could reflect the high level of personal acquaintance between the 1991 sample and the interviewer which may have reduced test inhibitions. The present sample, by contrast, had not met their interviewer previously.

A further point of contact between my two studies is over the precise relevant variables mentioned. Consistent with the 1991 study, both thickness and material (which equals metalness) were cited by the Howe *et al.* sample. Moreover, colour was studiously avoided. However, as Table 3.2 makes clear, material was better understood than thickness, a finding that is perhaps more readily related to the background literature than to my 1991 results. Also diverging somewhat from the 1991 data is an understanding of surface area that compares well with thickness. Surface area was referred to in the 1991 study. Thirteen children said a wide pan would heat the water more quickly and three identified a thick fork as protection from burning. This made surface area respectively the third and fifth most frequently used variable. Nevertheless, this is a far cry from the frequencies associated, as explained earlier, with thickness. It seems fairly obvious that such discrepancies reflect the more directive methodology of the Howe *et al.* study, a point that certainly needs to be borne in mind when considering how variable knowledge is structured.

MECHANISMS OF HEAT TRANSFER

The issue of structuring that concerns us most is the extent to which beliefs about variables are derived from causal mechanisms. As mentioned earlier, there are three mechanisms by which heat energy is transferred: conduction, convection and radiation. Conduction is heat transfer through solids and occurs because heating causes the constituent atoms (and electrons) to vibrate at relatively large amplitudes. The increased amplitudes are passed from atom to atom during collision between adjacent atoms. Heat transfer by convection occurs when a fluid is in contact with a heat source. The fluid that is in closest contact will increase in temperature and by virtue of this expand in volume. Becoming as a consequence less dense than the cooler surrounding fluid, it will rise.[1] The surrounding fluid will rush in to take the risen fluid's place and will in turn also be warmed and rise. This leads to a 'convective circulation' which over time warms the whole fluid. Radiation involves the emission of electromagnetic waves through low density media like air. Noting the existence of conduction, convection and radiation, the present section will use them as a yardstick for thinking about children. It will come as no surprise to learn that very few children acquire full understanding prior to teaching, but degrees of partiality will emerge which have interesting properties.

Separation of hotness from coldness

The first sign that children have problems with the heat transfer mechanisms has already been hinted at in our discussion of Clough and Driver (1985a). As intimated earlier in the present chapter and in part I, Clough and Driver found pupils talking about metal as a material that readily lets in both heat and cold. Thus, metal spoons feel hot when they are immersed in hot water because metal lets in the heat. The metal part of a bicycle handle feels cold after a night out of doors because metal also lets in the cold. Similar ideas are cited by Erickson (1979) and by Tiberghien (1980) in a study of French pupils aged 12 to 13 who were working on classroom practicals. For instance, it was not unusual in Tiberghien's research to find remarks like 'Metal cools things, metal is cold' going hand-in-hand with remarks like 'I've been told that metal heats up faster than any of the other three.' What these comments signal is the suggestion that heat and cold exist for children as independent phenomena. If this is the case there seems no chance of an adequate conception of conduction, convection and radiation. Adequate understanding requires a unitary notion of heat energy, a notion which involves movement from a relatively hot environment to a relatively cold one. It has no space for differentiation into hotness and coldness.

Unfortunately the separation of hotness from coldness is indicated by research quite apart from that relying on overt comments. It is implied, for instance, by three studies which were concerned not with causal mechanisms but with the consequences of mixing water at specified temperatures. The earliest and best known of these studies was conducted by Strauss and Stavy (and reported by Stavy and Berkowitz, 1980). Some 200 children aged 3-and-a-half to 14-and-a-half were presented in individual interviews with four types of problem: (a) qualitative/same, for example what happens when hot water is added to hot; (b) qualitative/different, that is what happens when hot water is added to cold; (c) quantitative/same, for example what happens when water at 70°C is added to water at 70°C; (d) quantitative/different, for example, what happens when water at 70°C is added to water at 30°C? Following on from Strauss and Stavy is the work of Appleton (1984) and Driver and Russell (1982), which in both cases included equivalent items as part of wider research. In the case of Appleton, the items were presented in interviews to twenty-five pupils aged 8 to 11. In the case of Driver and Russell, the context was a written survey with British and Malaysian pupils, with over one hundred participating at each of ages 8 to 9, 11 to 12 and 13 to 14. The results across the three studies were entirely consistent: quantitative problems were harder than qualitative but regardless of this, there was also a tendency for polarisation in the 'same' conditions. Hot plus hot produced very hot, and cold plus cold very cold.

It is difficult to interpret the polarisation without invoking two independent phenomena, 'hotness' and 'coldness'. Granted hotness and coldness, children would seem likely to reason that when two volumes of equally hot water are mixed there is more hotness and hence the temperature is higher. When two volumes of equally cold water are mixed, there is more coldness and the temperature is lower. Differentiation is suggested, then, by research which focuses on the mixing of water, but this is not all. There is further evidence in a somewhat different body of work. For example, in addition to replicating Strauss and Stacy, Appleton showed his respondents a picture of large and small ice cubes. Twelve of the twenty-five said that the larger cube would take longer to melt because it was colder. Likewise when shown a jug and a cup of hot water, ten said that the cup would cool quicker because it was colder. This was despite the fact that the water in the cup was visibly removed from the jug! Equivalently in Driver and Russell's study, 50 per cent of the 8- to 9-year-olds thought that the larger of two blocks of ice would be colder than the smaller, and 70 per cent thought that the temperature of a large volume of boiling water would be higher than that of a small. Similar ideas were not unknown amongst 13- to 14-year-olds. Once more, it is hard to interpret such findings without referring to two phenomena, heat and cold.

Energy vs. fluid models

Accepting then that we are dealing, at the very best, with the conduction, convection and radiation of two separate phenomena, the next question must be whether the phenomena are construed in a fashion that resembles 'energy'. Research by Gair and Stancliffe (1988), Solomon (1983) and Watts (1983) should make us pessimistic. Gair and Stancliffe interviewed fifty-three pupils aged 11 to 12 about the application of the words 'force' and 'energy' to contexts involving toys and simple household objects. These included a windmill, a matchbox car, a torch and a candle. Three frameworks were identified for both 'force' and 'energy', two 'animistic' and one primitively mechanical. Specifically, with energy, about 60 per cent of the sample limited the idea either to living things performing some action, for example 'Oh you have energy in games lessons' (Gair and Stancliffe, 1988: 171) or to non-living things imbued with human properties, for example 'Like, kind of the wind's energy. It's like strong and that. . . . Like the wind blows them. Sometimes it's strong' (Gair and Stancliffe, 1988: 172). The 40 per cent who displayed some mechanical understanding saw energy as the ability to produce an action or effect in a person or working object. This often went hand-in-hand with the acknowledgement that energy could be stored. However, the recognition of storage appears to have resulted in the differentiation of heat from the concept of energy rather than its incorporation within this. Thus, one pupil quoted in depth made comments like 'That is energy and it's just like storing it and with the heat it like melts away . . . when the flame burns, it uses the stored energy up, like the wax' (Gair and Stancliffe, 1988: 177).

Gair and Stancliffe's results find parallels in the work of both Solomon and Watts. Solomon asked 914 pupils to write sentences showing how they would use the word 'energy'. The pupils were described as being in the first three years of secondary school, and hence must have been aged between 11 and 14. Over 80 per cent of the youngest pupils produced sentences with living associations and although the percentage decreased with age it was never inconsequential. Interestingly, both Gair and Stancliffe and Solomon found that at all age levels girls were more likely than boys to equate energy with life. While not addressing gender nor giving quantitative data, Watts confirms the general picture. Based on interviews with forty pupils aged 14 to 18, he finds 'energy' being used in both a 'human-centred' and an agentive fashion. Thus the overall message across the research is conceptions of energy in the middle school years that make it highly unlikely that heat (or, for that matter, cold) is included. It will therefore come as no surprise to hear that research which has approached the issue directly finds the notion of heat energy to be extremely elusive.

Typical of the research is a study by Erickson (1980) where 276 pupils aged 11 to 15 watched a series of demonstrations, for example the melting on a hot plate of various substances. The pupils were asked to respond to a 'Conceptual Profile Inventory' by indicating which of several explanations of heating they most agreed with. The received 'kinetic' view which calls upon energy was always available for selection. However contrasting with this was a range of more 'childish' views, reflecting what Erickson (1979) had observed in the informal investigation mentioned in the previous section. Prominent amongst the childish views was the idea that heat is a fluid, an idea which according to Erickson (see also Erickson and Tiberghien, 1985) has parallels with the 'caloric theory of heat' that prevailed amongst scientists during the eighteenth and nineteenth centuries. Interpretation of Erickson's results is not entirely straightforward because nearly 50 per cent of his sample appear to have given inconsistent responses and were not as a consequence classified. Amongst the remainder, only 39 per cent subscribed to an unmistakably kinetic view, with most of these being at least 13 years of age.

By presenting competing explanations in a multiple choice format, Erickson may have overestimated his sample's knowledge. Childish ideas are after all often associated with childish words and thus stylistic features may have hinted at the 'correct' answers. As a result, it is useful to refer also to some work by Shayer and Wylam (1981) which invited pupils to write their own explanations rather than choose from a list. Essentially Shayer and Wylam combined Erickson's demonstrations with three that featured in research reported by Piaget (1974). The latter included a steel ball which was warmed and plunged into cold water, and wax affixed to rods of varying material which melted at differential rates when the rods were heated. Some 161 pupils observed the demonstrations and wrote explanations, with the pupils coming from two classes of 9- to 10-year-olds, one class of 11- to 12- year-olds and three classes of 12- to 13-year-olds. Shayer and Wylam report that 'Not more than one in five of the twenty three eleven- to thir-teen-year-olds who produced concepts at the above ("early formal") level offered kinetic theory models of conduction, heat transfer or gas expansion, and only one in two thought of expansion of solids in terms of greater movement of particles. Conduction was usually treated in terms of a "heat fluid caloric model" ' (Shayer and Wylam, 1981: 431).

It is interesting that Shayer and Wylam write as if conduction can be conceptualised independently of a kinetic theory. If on the one hand children differentiate heat from cold and on the other regard them both as fluids, it is hard from the received science perspective to see how they could be viewed as subscribing to conduction. Nevertheless, Shayer and Wylam probably have a point. In differentiating conduction from the theory that it belongs to, they are highlighting the possibility of distinguishing the means of heat transfer from the nature of what is transferred. It is conceivable that children see heat/cold as travelling gradually through solids in accordance with conduction, even if their images of what is travelling centre on fluids. It is likewise for convection and radiation where a vague notion of how transfer takes place could exist independently of any understanding of what is transferred. Is this what happens in practice? Unfortunately, virtually all the relevant research has focused on conduction. For convection and radiation, we have to be content with isolated quotations such as 'Heat comes from the radiator, it's like smoke for example, that comes and invades the whole house' (cited by Tiberghien, 1980: 85) and 'Most heat travels through some kind of rays' (cited by Erickson and Tiberghien, 1985: 60). There is no evidence regarding the frequency of such remarks nor regarding what else children say in similar contexts.

For conduction, there is also a range of choice quotations, for instance 'Metal is a conductor, it conducts heat up into the metal . . . it transfers heat . . . transfers heat along it' (from Tiberghien, 1980: 84) and 'The heat keeps moving from one point of the rod to the next until the whole rod is hot' (from Erickson and Tiberghien, 1985: 58). In addition however, the literature on conduction presents frequency information and, perhaps more significantly, also hints of developmental change. The research relating to the latter started with Piaget (1974). Using the three demonstrations later deployed by Shayer and Wylam, Piaget asked children to predict/observe outcomes and explain how they happened. He detected at least two levels of mechanism: (a) contagion, that is objects 'catch' heat from being in the environment of heat sources; and (b) transmission, that is objects become hot because heat travels through some medium up to them (in other words, the 'process' equivalent of conduction). While Piaget recorded contagion in children as young as 6, he claimed that transmission was rare until his 'Stage III' which began at 11 to 12 years. Shayer and Wylam seem to confirm the latter via items in their study which bore on 'the movement of heat'. As far as I can gather, the sense that heat travels gradually through things was limited to their most advanced group. Since there was a correlation of +0.62 (df = 159, p<0.001) between age and advancement in a sample that varied in age from 9 to 13, the results appear to parallel Piaget's exactly.

Developmental research into the transfer mechanisms

When I read the results of Piaget and Shayer and Wylam, I was both intrigued and suspicious. I was intrigued by the apparent existence via contagion of transfer mechanisms in addition to transmission. I was suspicious because transmission was being aligned with Stage III of Piaget (1974). As presented in the 1974 book, Stage III is the stage at which actions and objects are sufficiently differentiated to allow the possibilistic perspective on variables that was discussed in part I, the so-called 'formal operational stage' of Inhelder and Piaget (1958). Thus, as we saw in part I, it is most definitely the stage at which Piaget would anticipate improvements in children's understanding of mechanisms, meaning that his transmission data concur closely with his theory. However, there was to me an intuitive straightforwardness about the *concept* of transmission which made me wonder if it was really the late (and problematic) acquisition that was being suggested. I resolved to incorporate questions that would shed some light into the Howe (1991) study which was introduced in the previous section.

Thinking how to proceed, it seemed to me that there were two possible sequences of questioning. One would involve following questions about heat transfer outcomes with questions about how outcomes were achieved, for instance 'What will happen to the water once the cooker is switched on? How will that happen?' The other would involve following the questions pertaining to variables with 'Why is (variable) important?', for example 'Which pan would be best? Why? Why do you think that (having a metal pan) would help the water heat up?' The rules of conversation as outlined by Grice (1975) would lead adults to take the final questions in both sequences as requesting mechanisms, and I felt that the same was likely to apply with children, at least at the age level being studied. There is an extensive literature attesting to mastery of the Gricean rules by the first years of schooling (see Foster (1993) for a summary). Moreover, Dunn (1989) has presented evidence that school-age children will have known for some years about justifying themselves, and indeed justifying their justifications, in response to adult questioning.

The children's competence recognised, it seemed to me that the first of the two possible sequences of questioning could be accused of bias towards responses which differentiated mechanisms from variables, since mechanisms were being requested in contexts where variables were unlikely to be mentioned. Likewise, the second possible sequence could be accused of bias towards responses which linked mechanisms and variables, since questions geared at the former followed and were integrated with questions geared at the latter. The former bias would favour the action-based approaches of the theories under test; the latter the theory-based. Since the goal had to be a research design which was neutral between the approaches, I decided that the only solution was to include both sequences, and this is how the 1991 study proceeded.

As it turned out, the results were closely in line with the antecedent research. First, virtually all the pupils demonstrated an understanding that heat is transferred. It has to be pointed out that this understanding was more likely to be demonstrated with the pans scene than with the forks, for while only five of the 126 pupils failed to indicate that the cooker's heat would transfer to the pan, twenty-five failed to indicate that the fire's heat would transfer to the fork. With the pans scene, four of the five 'failures' indicated a non-transferring mechanism, for example 'The heat's inside the water'. With the forks scene, only one of the twenty-five did this, the remainder not giving any mechanisms at all. Since the forks scene was always presented immediately after the pans, this suggests that in many cases the failure to acknowledge transference reflected the desire not to repeat rather than anything more fundamental.

In addition though (and more importantly) the mechanisms of transfer that were called upon were not invariably transmissive. A total of 101 pupils responded to the pans scene with remarks like 'It heats it from the rings with electricity' or 'The heat's underneath it.' Fifty responded equivalently to the forks scene with the likes of 'The heat's coming from where the fire is' or 'The long fork will keep her away from the flames.' Whether these responses are adequately captured by Piaget's (1974) concept of 'contagion' seems debatable, but certainly there is no recognition of transmission in his sense. The remaining children did however give transmissive replies, via such remarks as 'The heat of the stove will come up through the pot and heat the water' and 'The heat will go up the metal and burn'. As these examples suggest, transmission relatable to conduction predominated in the responses and this is not altogether surprising. Both the pans scene and the forks scene depicted conduction, while only the pans depicted convection and neither depicted radiation. What is perhaps more surprising is that transmission relatable to convection appeared on only two occasions in total. One of these involved the claim that 'The hot water at the bottom rises in small bubbles.'

Thus, as Piaget suggested, transfer appears to be common but transmission seems much rarer. How, then, about age trends, the third strand to Piaget's argument? Here too the data look consistent. As Table 3.3 shows, there was a decrease with age in the number failing to recognise transfer and an increase with age in the number acknowledging transmission. This led to statistically significant age effects for mechanism with both the pans scene (X^2 = 17.78, df = 8, p<0.05) and the forks (X^2 = 45.77; df = 8, p<0.001). What is particularly interesting in the light of Piaget, and Shayer and Wylam and of course the theoretical issues being researched in this book is the clear division in Table 3.3 between the two oldest groups and the others. Indeed, for the pans scene though not so much for the forks, there is virtually no difference between any of the three youngest groups.

To consolidate the evidence still further, I looked in addition at the data collected for the Howe et al. (1995a) study that was also reported in the previous section. As part of the investigation described in Howe et al., participating pupils were asked why the variables they cited were important,

Table 3.3   Mechanisms of heating as a function of age: temperature change (Howe, 1991)

Pans	6–7- year-olds	8–9- year-olds	10–11- year-olds	12–13- year-olds	14–15- year-olds
No transfer	4	1	0	0	0
Transfer but no transmission	20	22	21	17	21
Transmission	2	2	3	7	6
Forks	*6–7- year-olds*	*8–9- year-olds*	*10–11- year-olds*	*12–13- year-olds*	*14–15- year-olds*
No transfer	14	7	4	0	0
Transfer but no transmission	11	12	9	10	8
Transmission	1	6	11	14	19

and the responses they gave were coded for knowledge of mechanisms. Coding proceeded in a more quantitative fashion than was the case in my 1991 study. In particular, the responses to each of the twelve interview items were scored on a scale which ran from zero to three. Responses which did not indicate a mechanism obtained a score of zero; responses which indicated a mechanism without knowledge of transfer obtained a score of one; responses which indicated transfer without knowledge of transmission obtained a score of two; and responses which indicated transmission obtained a score of three. Mean scores across items were computed for each pupil in the sample. Although arguably less revealing than the item-by-item breakdown in my 1991 study, it is instructive that here too there was an age-related improvement in the pupils' scores (F = 10.83, df = 3,96, p<0.001). Scheffé tests revealed that the 11- to 12-year-olds (mean = 0.76) obtained significantly higher scores than the 8- to 9-year-olds (mean = 0.19), the 9- to 10-year-olds (mean = 0.24) and the 10- to 11-year-olds (mean = 0.44). No other age groups differed significantly.

With the Howe *et al.* (1995a) data, we then have evidence which, like the Howe (1991) study, points to an age-related improvement in pupils' understanding of the heating mechanisms. However, we do not only have this. We also have evidence for an improvement that is located just prior to the teenage years, after a long period of negligible change.[2] As a result, we have evidence that squares not just with the Howe (1991) data on heating mechanisms but also with the data from both studies on variable selection. As demonstrated earlier, there was in both cases an age-related improvement in pupils' subscription to relevant variables, after an extended period of no change. The improvement can be characterised as a 'big leap forward' between 10 and 12. As noted already, improvement in understanding of mechanisms and variables around but not before the age of 11 is more consistent with action-based approaches to knowledge than it is with theory-based ones. Moreover, within the action-based framework, it concurs better with the Piagetian stance than it does with the Vygotskyan. Given the nature of mature everyday physics, improvement is predicted on the Vygotskyan model throughout the school years and not simply after 11. However, supporting Piaget implies denying theoretical structure before the age of 11. Is this really to be countenanced? In the next section we shall attempt to find out.

TRANSFER, TRANSMISSION AND VARIABLE SELECTION

If we ignored the spurt between 10 and 12 in mechanism and variable knowledge and its apparent consistency with Piaget, we should be encouraged by the temporal association between the two to expect a contingent relation and perhaps even the dependency of variables on mechanisms. Could such contingency still exist? Could there even be dependency? The present section will start by presenting data which bear on these questions.

Obtaining evidence which is indicative of both contingency and dependency, the section will then discuss the paradox of support for Piaget from the knowledge spurt and rejection from the structural relation. The points raised will form a backcloth to further investigation in the chapter to follow.

Patterns of association

As a first step towards exploring contingency and dependency, I calculated the correlations between the numbers of relevant and irrelevant variables that each pupil mentioned in my 1991 study and what I assigned as their 'mechanism score'. The latter was a value from zero to two for each of the scenes, and depended quite simply on the Table 3.3 category. Assigning scores in this way seemed (more or less) justified given that the categories in Table 3.3 lie unambiguously on a continuum of adequacy. It is, at the very least, unlikely to be misleading. In any event, I proceeded and when I compared the values correlationally, an interesting pattern emerged. There were significant positive correlations between the numbers of relevant variables and the mechanism scores for both the pans scene and the forks (r for pans = +0.29, df = 124, p<0.001; r for forks = +0.49, df = 124, p<0.001). By contrast, the correlations between the numbers of irrelevant variables and the mechanism scores were in both cases insignificant.

Focusing then on relevant variables, the question is what do the correlations mean. Of particular interest is whether changes in mechanism knowledge could have been driving changes in relevant variables, or whether the reverse was more likely to have been the case. Without repeated observations of the participating pupils, there is a limit to what can be said here. Nevertheless, something can be established by looking at the proportion without transfer (and then transmission) who pass some threshold of relevant usage compared with the proportion with transfer (and then transmission) who fail to pass the threshold. To the extent that the former is smaller than the latter, expansion of relevant variables is more likely to wait on mechanism change than for the reverse to be the case. With this in mind, the pupils were tabulated as shown in Table 3.4.

From Table 3.4, it is clear that the proportion of pupils reaching given levels of mechanism knowledge without corresponding increases in relevant variables was consistently greater than the proportion reaching given levels of relevant variables without corresponding increases in mechanism knowledge. For example, the percentage of pupils who had passed the 'no awareness of transfer' stage but did not mention any relevant variables was 50 per cent for the pans scene and 11 per cent for the forks. The percentage who mentioned at least one relevant variable but showed no awareness of transfer was 0 per cent for the pans scene and 9 per cent for the forks. The percentage of pupils who understood that heat is transmitted but mentioned no more than one relevant variable was 14 per cent for the pans scene and 22 per cent for the forks. The percentage who mentioned two relevant variables

Table 3.4 Relation between mechanisms and relevant variables: temperature change (Howe, 1991)

Pans	0 Variable	1 Variable	2 Variables
No transfer	5	0	0
Transfer without transmission	57	42	2
Transmission	6	12	2
Forks	0 Variable	1 Variable	2 Variables
No transfer	14	11	0
Transfer without transmission	9	34	7
Transmission	5	23	23

but did not appreciate that heat is transmitted was 2 per cent for the pans scene and 6 per cent for the forks.

Wells (1985) has described a technique known as the Critical Ratio Test which indicates the statistical significance of distributions such as the above. The test requires the data to be cast into two-by-two contingency tables, and involves the calculation of z values.[3] Significance at the 0.05 level requires z to be at least +1.96 or -1.96, with the meaning of positivity and negativity depending on how the tables are drawn up. For present purposes, z values were computed such that positivity meant that mechanism 'gains' without variables were more likely than the reverse. The results for awareness vs. non-awareness of transfer and zero vs. one or more relevancies were +7.93 (p<0.001) for pans and +0.60 (ns) for forks. The results for awareness vs. non-awareness of transmission and one or fewer vs. two relevancies were +3.58 (p<0.001) for pans and +3.55 (p<0.001) for forks.

Obviously, the results just presented do not prove that mechanisms were playing a generative role in variable selection, but nevertheless they do enhance the probability that this was going on. Moreover, they enhance the probability that mechanisms were generative at all ages considered, for the relations applied to the transfer mechanism that was ubiquitous across the sample as well as to the transmission mechanism that increased in frequency after 10 or 11. This 'age-independence' was confirmed when the analyses were repeated for the fifty-one pupils in the two youngest age groups, and then for the seventy-five pupils in the three oldest. (This cut-off was selected because the 'big leap forward' aligned the 10- to 11-year-olds with the 12- to 13- and 14- to 15-year-olds and not with the younger age groups.) With the pans scene, the correlations between numbers of relevant variables and mechanism scores remained positive but did not reach statistical significance (r for younger pupils = +0.15, d = 49, ns; r for older pupils = +0.20, df = 73, ns). With the forks scene, the correlations were positive and, in both cases, statistically significant (r for younger pupils = +0.43, df = 49, p<0.001; r for older pupils = +0.30, df = 73, p<0.01). Some 22 per cent of the younger pupils reached the various mechanism levels without passing the relevancy thresholds as opposed to 17 per cent who did the reverse. The corresponding figures for the older pupils were 35 per cent and 12 per cent respectively.

Bearing these results in mind, we can perhaps move onto the Howe *et al.* (1995a) data. There, it will be remembered, variable selection was quantified by assigning four scores of zero to three to the responses that each pupil made to each of twelve items. The scores corresponded to knowledge along the thickness, material, surface area and colour dimensions. Five mean scores were then computed for each pupil across the items, the means reflecting the four dimensions taken separately plus all four combined. All but knowledge of colour improved with age, although in the case of surface area the improvement was not statistically significant. As we saw in the preceding section, mechanism knowledge was also quantified, by assigning scores of zero to three to the responses to each item and computing means for each pupil across the item set. Thus, with both variables and mechanisms, we have data where the computation of correlations can be treated as uncontroversial. Taking this for granted, I computed five correlations, that is between mechanism means and each variable mean in turn. With the exception of colour, all correlations were highly significant (r for thickness = +0.54, df = 98, p<0.001; r for material = +0.35, df = 98, p<0.001; r for surface area = +0.36, df = 98, p<0.001; r for combined = +0.63, df = 98, p<0.001).

Change over time

While the data collected for the Howe *et al.* (1995a) study lend themselves uncontroversially to correlational analysis, they square less readily with breakdown as signified by Table 3.4. Where they compensate however is in being part of a wider dataset which bears incisively on the processes by which variables and mechanism change. The raison d'être for the Howe *et al.* study was not to collect the descriptive information that we have taken from it so far but rather to test the utility of a particular approach to teaching. This approach was inspired by two considerations. The first was the evidence (summarised by Howe, 1995) that knowledge of variables can be promoted by group tasks where pupils have to formulate joint predictions. The reason for this is that pupils call upon variables to resolve prediction disagreements (just as in the interview studies reported in this chapter, they called upon variables to justify decisions about facilitators of outcome). The second consideration was the success of the Howe (1991) procedures in getting children to use mechanisms to justify variables. Putting the two considerations together, the way forward as regards teaching variables and mechanisms simultaneously seemed to be group tasks where pupils do not simply formulate predictions but also reflect on and appraise the variables they generate. The Howe *et al.* study was designed to investigate whether this was correct.

Following an earlier study by Tolmie *et al.* (1993), a procedure called 'rule generation' seemed crucial to supporting variable appraisal. What this means is taking groups through cycles of activity where they formulate joint predictions, test the predictions empirically, interpret jointly what transpires and then, with reference to all cycles completed, decide 'What is important' in the domain of interest. However, experience had shown that unfettered predict–test–interpret cycles typically produce such a vast array of variables that pupils can find the rule generation unmanageable. Accordingly, it seemed wise to constrain the variables by presenting test items in ordered sets, such that they differed with respect to one variable only. It was anticipated that this would focus the group's attention, limiting their discussion and perhaps even allowing them to subject variables to 'critical tests'. The Howe *et al.* study was in essence an appraisal of rule generation plus critical testing in the context of heat transfer.

To this end, the study had a three-stage design, beginning with a 'pre-test' to establish baselines prior to the group tasks. It is the pre-test data that have been presented thus far. Soon afterwards ninety-six of the pre-tested pupils were assigned to groups of four to work collaboratively on the crucial tasks. For all groups, the tasks involved filling pairs of containers with equal volumes of freshly warmed water, and predicting privately (in writing) the container in which the water would cool down the fastest. Subsequently, group members were invited to share their predictions, resolve disagreements, use a digital thermometer (which they had been trained to operate) to ascertain whether their agreed predictions were correct, and jointly interpret outcomes. Group interaction was recorded on videotape. The third and final stage took place between three and eight weeks after the tasks, when group participants were individually 'post-tested' along the lines of pre-test and progress was quantified with reference to pre- to post-test change.

Within the basic task structure that was outlined above, there were four distinct formats: rule generation plus critical test, rule generation alone, critical test alone and 'random' (that is neither rule generation nor critical test). With the two critical test formats, the containers were presented in ordered pairs such that one of thickness, material, surface area and colour varied while the others were held constant. With the two rule generation formats, conclusion of certain predict–test–interpret cycles was accompanied by an instruction to discuss and write down 'things which are important to how quickly hot water cools down'. Six groups worked with each task format. The expectation was that the rule generation plus critical test format would promote productive discussion of mechanisms and by virtue of this facilitate mechanism change. The results supported this: there was a significant difference in pre- to post-test mechanism change as a function of task format ($F = 9.96$, df $= 1,84$[4], $p<0.01$) with the greatest mean change associated with rule generation plus critical test. Knowledge of variables also advanced from pre- to post-test, but this time there were no differences as a function of task format. This should not be surprising given that all groups experienced the predict–test–interpret structure which had already been established as productive with respect to variables.

The overall success of the rule generation plus critical test format is important. However, of greater consequence here is the process by which success was achieved, a process that was observed with all age groups studied. As expected, analysis of the videotapes revealed that disagreements over variables which occurred during the rule generation exercise were likely to be resolved with reference to mechanisms. This proved to be crucial since the calling on mechanisms in the rule generation context had an immediate positive influence on the mechanisms themselves, for these were likely to be what Howe *et al.* called 'high level'. High-level mechanisms are equivalent to what we know as transfer by transmission (and thus are 'high level' in only the most relativistic sense). The high-level mechanisms produced in response to rule generation were subsequently referred back to when, on later task items, disagreements arose at the prediction stage. Thus, typical of the rule generation plus critical test groups were the following dialogues:

Rule generation

LEE-ANNE   I think the thickness of the sides.
GAVIN   Cos it'll keep more heat from going out.
KEVIN   But the tops are smaller, and that'll contain more heat.
LEE-ANNE   The tops are smaller and there's more thickness.

Prediction disagreement

LEE-ANNE   I say the black one.
GAVIN   I say the grey.
LAURA   I say both of them.
KEVIN   And I say both of them.
LAURA   They're both exactly the same except that they're different colours.
LEE-ANNE   So they must let exactly the same amount of heat pass through.

Such reference back to mechanisms during prediction may have helped 'cement' the newly formulated mechanisms in the pupils' minds. Certainly, they must have been cemented because pre- to post-test change in mechanism scores was correlated +0.58 (df = 20, p<0.01) with references during the group task to high-level mechanisms. Indeed, pre- to post-test change in mechanism scores was significantly correlated with high-level mechanisms, disagreements over variables, references back to mechanisms and disagreements over predictions, and these factors were all strongly associated with each other (r = +0.89 to +1.00, df = 4, p<0.01). Neither pre- to post-test mechanism change nor the critical aspects of dialogue were correlated with any other aspects of group interaction.

The point is then, that on-task variable discussion produced on-task mechanism change which was preserved over time. However, what can be said about changes in the variables themselves? Howe *et al.* discuss changes in variable knowledge but they focus on an index of knowledge which is different from, though related to, the ones that we have been using here. Analyses of pre- to post-test change have also been carried out on our indices and they mirror exactly what Howe *et al.* report. This is that pre- to post-test variable change was strongly associated with pre- to post-test mechanism change, being correlated to the extent of $+0.54$ (df $= 20$, $p<0.01$) in the case of the measure which combines across our dimensions. However, pre- to post-test variable change was unrelated to the dialogue factors that predicted mechanism change, nor indeed to any other parameters of group interaction. Faced with these results, it is hard to escape the conclusion that mechanism change was making the running. In other words, to complete the story, on-task discussion of variables precipitated on-task progress over mechanisms. This was preserved over time, and fed as a result into knowledge of variables.

Looking at the results with the rule generation plus critical test format, two interpretations come to mind. The first is that change took place in the way that it did because of constraints imposed by the pupils' knowledge structures. If this interpretation is correct, it would mean that mechanisms are generative in the context of heat transfer, and children's knowledge can legitimately be described as theoretical. The second interpretation is that change took place in the way that it did because rule generation plus critical test restricted the options.

As it happens, the second of the two possible interpretations is open to two objections. First, the point of rule generation plus critical test was to create dialogue where predictions are justified by variables and variables are justified by mechanisms. It is not apparent a priori why such dialogue should have had an immediate impact on knowledge of mechanisms but not on knowledge of variables. Second, there were signs that the other task formats were *preventing* change from taking place in the manner of rule generation plus critical test rather than substituting something different. Critical testing alone served to reduce the number of variables considered to no productive end. Rule generation alone established a context where mechanisms were used to arbitrate between variables, but these mechanisms were as the following extract shows very low level:

Rule generation

IAN   It's cos they're wider.
STEPHEN Cos they're thick and narrow. It comes in more, and less air can get in. That's small and wide, so the air pressure can get in. Don't you understand that?

Needless to say, the on-task mechanisms associated with rule generation alone were negatively correlated (r $= -0.47$, df $= 20$, $p<0.05$) with mechanism change. Finally, there was no reason to see the random task format as operating in a different way from what was established for this format by Tolmie *et al.* (1993). This was that the random structure leads to the generation of a large pool of variables which only proves helpful after a long period of post-task reflection. In the present context, it is tempting to suggest that the post-task reflection was necessitated because variables were generated apart from mechanisms and could not therefore be co-ordinated with existing knowledge.

Preliminary conclusions

Obviously, any conclusions from the Howe *et al.* (1995) results must be tempered by the fact that the study is an isolated piece of work where a series of investigations would be preferred. Nevertheless, the picture created by the processes of change is consistent with what the earlier correlational analyses suggested. This is that for the contexts of heat transfer studied so far, mechanisms appear to be playing a generative role. Certainly, the generative role is limited to the variables that science deems relevant. The Howe *et al.* study looked only at relevancies, and the Howe (1991) study failed to find contingency between irrelevant variables and mechanisms. However, it is interesting that in the 1991 study, usage of irrelevant variables was relatively uncommon, and was in any event negatively correlated with usage of relevancies. From the rarity, it might be possible to argue that irrelevancies were 'random noise' brought on perhaps by demand characteristics of the interviews. From the negative correlations, it could be suggested that irrelevancies did in fact stem from mechanisms but in an indirect fashion. In particular, improvements in mechanisms led to the generation of relevant variables, and incompatibilities between relevant variables and the pre-existing irrelevancies led to the elimination of the latter.

The proposal that irrelevancies 'pre-existed' the operation of mechanisms hints strongly of the action-based approaches to conceptual growth. Thus, in its own right, it continues the support for the Piagetian version of those approaches that came earlier from the leap in variable and mechanisms understanding around the age of 11 after a very slow start. This time though, it also squares with Vygotsky. However, the alternative line, that the irrelevancies were nothing more than noise, should not be overlooked. If that line was correct, it would be possible to argue that within conceptual structure nothing of significance lies outside the scope of mechanisms, and this of course would signal the theory-based approaches.

Up to a point, this consistency with the theory-based approaches would be all of a piece. No matter what gloss is placed on irrelevancies, the fact remains that relevant variables were theoretically generated, and were so from a very early age. After all, the results reported in the past few paragraphs were as true of the 7-, 8- and 9-year-olds as they were of the older age groups. This is exactly what the theory-based approaches would anticipate, although Vygotsky could also deal with it. Only Piaget despite all the other support for his thesis would be left high and dry. So what do we make of this? On the one hand, we have evidence for a leap in variable and mechanism knowledge around 11 after a lengthy period of stasis, and only Piaget could deal with this. On the other hand, we have evidence for early theorising which Piaget would deny but which both the theory-based approaches and Vygotsky would feel comfortable with. We have a dilemma. However, at this stage, we should probably not try to resolve it. Rather we should hold fire, and seek further clarification with other aspects of physics. The chapter to follow will make a start by considering a contrasting aspect of the transfer of heat.

4 The 'peripheral' case of changes of phase

With the material discussed in the previous chapter, we seem to have identified an area where children's knowledge shows a stepwise progression in content but is theoretically structured from the start. However, is this true with all contexts where heat is transferred? Up to now, we have restricted our attention to contexts where the transfer of heat implies a rise in temperature. However, during changes of phase-like melting and solidifying, evaporating and condensing, heat is transferred but temperature remains constant. Yet the variables which science deems relevant to changes of phase and the mechanisms that are called upon are equivalent to the ones identified in the previous chapter for temperature change.

How then do children deal with contexts involving changes of phase? Do they continue to rely on a small set of variables, most of which are scientifically relevant? Does the number of relevancies continue to improve with age? Do the variables continue to be associated with causal mechanisms in a fashion that appears to be generative? Does the nature of the causal mechanisms continue to progress with age despite no change in the generative relation? Finally, does improvement in variables and mechanisms continue to mean a big leap forward around 11 or 12 after a virtually static situation? In the present chapter, we shall attempt to answer all of these questions. As we proceed, we shall unearth a situation which re-echoes the previous chapter's in many respects. Nevertheless, while this is the case, there is also an element of 'fuzziness' with changes of phase which was not observed earlier. Age-related advances occur somewhat later than they did with temperature change and the demarcation between relevant and irrelevant variables is less clear-cut. This will lead to the conclusion that events involving changes of phase are 'peripheral' within a concept where events involving temperature change are central.

VARIABLES RELEVANT TO PHASE CHANGE

On the face of it, there is every reason to think that children's beliefs about changes of phase will be linked to their beliefs about changes of temperature. In the first place, everyday experiences of changes of phase centre on events where hitherto there has been marked temperature change. The water in the saucepan typically heats up before it evaporates into steam. The water on the pond (not to mention the ambient air) typically cools down before there is freezing into ice. In addition though, there is a series of published studies which indicates that many children are unaware that temperature remains constant during changes of phase. On the contrary, they see boiling as continuing the process of temperature rise initiated by heating, and freezing as continuing the process of temperature fall initiated by cooling. The present section will start by reviewing these studies, so that their results may be borne in mind when it moves to its main theme of attributed variables.

Phase change as a conceptual distinction

The studies which address beliefs about temperature during changes of phase all involve large numbers of school pupils who were given paper-and-pencil tests. One of the studies was carried out in Sweden by Andersson (1980). Andersson found that 105 out of 441 12- to 15-year-olds denied that water boiling in a saucepan would remain at the same temperature after five minutes. A total of 257 denied that it would do this after turning the cooker up higher. Similarly, in the survey with 900 British pupils aged 11 to 15 years that has been mentioned in previous chapters, Brook *et al.* (1984) observed several pupils claiming that potatoes would take longer to cook in slow boiling water than they would in fast because of the former's lower temperature. Beyond this, even at the oldest age levels, only 39 per cent could identify melting point on a temperature graph. Supplementing this is the study by Driver and Russell (1982) that was also referred to earlier. This study, it will be remembered, involved British and Malaysian pupils at three age levels: 8 to 9, 11 to 12 and 13 to 14. It turned out that only 10 per cent were aware of temperature constancy during melting, though somewhat more responded correctly to boiling. Finally, there is the work of Tiberghien (1984) where only 20 per cent of French pupils aged 10 to 13 knew that temperature is stable during boiling.

At first sight, the message seems to be that situations involving changes of phase are no different for many children from situations involving changes of temperature, that in other words phase change is not in many cases a bona fide conceptual distinction. However, two potential problems need to be noted. Firstly, in all of the studies, pupils were asked about temperature rise, not about becoming hot or heating up. This would not matter if they have an adult conception of temperature, but research reviewed by both Tiberghien (1984) and Wiser and Kipman (1988) suggests that this may not be the case. According to these writers, temperature is sometimes treated by children as a synonym of heat, sometimes as the term for the hot/cold dimension and sometimes as a measure of how much heat is present. In which case, the pupils in the studies may have been treating the question as concerned with heat flow, meaning that their answers were correct in scientific terms, but unrevealing about whether they extend heating up or cooling down to changes of phase. In addition however, the pupils in all of the studies apart from Driver and Russell's were at the older end of the range we are interested in. Another set of studies suggests that younger children may in some circumstances make a clear distinction between phase changes and the contexts we have hitherto been concerned with.

The studies that I am thinking about include ones that are concerned with evaporation. Amongst these, ironically enough, is the work of Driver and Russell (1982) who asked their respondents about water that had dried on a blackboard. However, they also include studies by Bar (1986, see also 1989), Beveridge (1985), Osborne and Cosgrove (1983) and Russell and Watt (1990). The important part of Bar's study is an interview administered to 165 5- to 11-year-olds which included questions about what would happen to water spilled on the ground. Beveridge's main focus was a training study, but we can draw some inferences from both his pilot interviews with seventy-six 7- and 8-year-olds and his control interviews with ninety 5-, 7- and 9-year-olds who did not participate in training. The topic of the interviews was water emanating from respectively puddles, drying clothes, ponds and a boiling pan. Along similar lines, Osborne and Cosgrove asked about water drying on a saucer, but this time to forty-three pupils aged 8 to 17. Finally, Russell and Watt asked sixty pupils whose ages spanned the primary school range about evaporation from a fishtank, a line of clothes, a coffee cup, a sugar solution and a paper towel.

The main finding from the studies is that pupils typically give one of three responses: the water just disappears; the water soaks into its surround; and the water goes into the air. Bar (1986, 1989) reports that the pupils who favoured disappearance tended to fail on Piaget's conservation of liquids task. Thus, here we should be talking of pupils who are typically under 7. This is what Bar herself found, although in the only other work with very young children, Russell and Watt obtained results that are less clear-cut. Bar further reports that recognition of the omnipresence of air (as opposed to its creation by motion) was a good predictor of saying that water goes into the air. Since as we shall see in Chapter 8, recognition is unusual below about 10, soaking-in responses should predominate amongst 7- to 10-year-olds. Here, Bar's results suggesting this have been amply confirmed by the other research.

The age norms are interesting but in one sense they are beside the point. The main point is surely that children who think water disappears or soaks in will not necessarily predict temperature change or call on heat transfer when dealing with evaporation. Indeed, the point should probably not be restricted to evaporation, for the work of Driver and Russell, Osborne and Cosgrove and in addition Piaget (1974) suggests extension to condensation and hence to liquid/gas changes in general. Driver and Russell instructed their sample to imagine the following scenario: 'We feel nothing on the outside of this empty beaker. We put cold water into the beaker. Some water collects on the outside of it.' They then asked 'Where does the water on the outside of the beaker come from?' Nearly 50 per cent of the youngest age group, the 8- to 9-year-olds, believed that the water had seeped through the beaker or come like rain from the air. Moreover, even at the 11- to 12-year age level, there were plenty of British pupils who were unaware that the water had formed by condensation. Interestingly, the Malaysian pupils of comparable age had better understanding. Osborne and Cosgrove along with Piaget provided pupils with demonstrations of the evaporation of water in a closed vessel followed by its reappearance through condensation. Many pupils saw no connection between the two events, frequently believing that the condensed water came 'from out there'.

Variables facilitating changes of phase

The purpose in discussing evaporation and condensation is not to deny that changes of phase will be aligned with changes of temperature in the thinking of children. Rather, it is to indicate that evidence pointing against alignment exists alongside evidence pointing towards it. In addition, it is to indicate that when considering variable selection and/or causal mechanisms in the contexts of evaporation or condensation, it might be advisable to take account of what children believe regarding where the liquid goes. Following the sequence of the previous section, we shall be discussing variable selection, then causal mechanisms and finally variable–mechanism relations in the paragraphs to follow. Thus, our first questions will include: (a) what variables do children regard as relevant to the rate at which phases change; (b) are these variables relatable to the ones that they call upon for changes of temperature; and (c) do beliefs about what is happening during evaporation and condensation influence variable selection?

Unfortunately while it would be gratifying to find a substantial body of evidence that bears on the issues, this cannot be done. The main thrust of research has been the variables that children call upon to decide whether phase changes take place, not the variables they deem relevant to how fast changes occur. Typical here is an investigation by Tiberghien and Barbaix (reported by Erickson and Tiberghien, 1985). Some 12-year-old French pupils were questioned about whether a range of solid substances could become liquid given a sufficiently high temperature. The majority thought that some substances could change while others could not, basing their decisions on observable properties such as hardness and/or specific experiences such as the need to melt gold to make gold bricks. Relatedly, Russell *et al.* (1991) asked sixty-eight pupils aged 5 to 11 what would happen to a metal rod, a piece of cotton wool and some vinegar if they were heated to a very high temperature. With the metal rod, a phase change, to wit melting, was anticipated by 48 per cent of the youngest age group and 86 per cent of the eldest. With the cotton wool, a phase change was never anticipated, for most children were confident in the knowledge that the substance would simply get hotter and/or burn. The results with vinegar replicated what we have already observed, a decrease with age in anticipated disappearance/soaking away and an increase in anticipated evaporation.

Probably the main point to be drawn here is that with the exception, once more, of evaporation, children are canny observers of when phase changes do and do not occur. Is this, then, as far as they go, or do they also make inferences about rates of change? As noted above, this issue has not been the subject of extensive research. In the interview study with twenty-five 8- to 11-year-olds that was outlined in the previous chapter, Appleton (1984) asked about the time that a large and small ice cube would take to melt. Twenty-three of the twenty-five pupils said that the small ice cube would melt the fastest. However, it is unclear from this whether they thought that large and small ice cubes melt at the same rate per unit volume, or that large ice cubes melt at a slower rate. Only in the latter case could size be interpreted as a bona fide variable. Noting this, it is of interest to recall what we saw in the previous chapter about the anticipated coldness of large ice cubes. This suggests (without proving) that children may indeed anticipate a slower unit rate of melting as size increases.

Also concerned with melting is the work of Erickson (1979) which, as mentioned previously, involved quizzing 6- to 13-year olds about classroom demonstrations. One demonstration involved the task with wax melting at differential rates from heated rods which as explained in the previous chapter was introduced into the literature by Piaget (1974). Data relating to this demonstration could have been extremely relevant but unfortunately Erickson does not organise his material in a helpful way. The aim of his 1979 paper was to present qualitative data regarding what children say (and not when they say it). Thus, it is sometimes difficult to decide whether given remarks were associated with the melting wax demonstration or with the other demonstrations which related instead to rises of temperature.

Summing up then, the published literature appears to be limited to two potentially relevant studies. Unfortunately, these studies focus exclusively on melting and are, in some respects, ambiguous. It was with these conclusions in mind that I incorporated three items tapping the variables relevant to phase change in the Howe (1991) study that was utilised so heavily in the previous chapter. The items related to photographed scenes involving: (a) four saucers on a shelf above a radiator, one holding a plant; (b) five wraps on a picnic rug, one being placed around ice cream; and (c) three wax candles alight on a dining room table. The saucers scene was used to quiz the pupils about evaporation, specifically the evaporation of water from the saucer which held the plant. Thus, cognisant of points raised earlier in this section, the interview schedule relating to this scene began with questions about where the water would be after a couple of days. After the pupils had answered such questions, they were asked 'Do you think the kind of saucer makes any difference to how long the water will last? Do you think the water will last longest in the saucer that is under the plant? Why? What about the other saucers? Will the water last about the same time in them, or will there be differences? Why?'

The wraps and the candles scenes provided contexts for enquiring about melting. In the case of the wraps scene, the attendant questions were fairly general, that is 'Do you think that wrapping something around ice cream will make any difference to how quickly it melts? What difference will it make? Do you think that some wraps are better than others at keeping ice cream from melting? Has the picnicker chosen the best wrap? Why? What about the other things? Would they all be equally bad or would some be worse than others?' In the case of the candles scene, the questions started general but became specific, that is 'Suppose that there was a power cut and you wanted the candles to last as long as possible. Is there anything that you could do to slow the melting down? What? Why? Would putting something over the candles make any difference? Why? Would moving them to another part of the room? Why? Would it make any difference to make the room cooler? Why?'

As with the pans and forks scenes discussed in the previous chapter, some pupils denied that changing the saucers, wraps or candles set up would be of relevance. However while only thirty offered denials with the wraps scene and only thirty-three with the candles, sixty-seven did this with the saucers. The discrepancy was almost certainly a direct consequence of the 'emergent' nature of evaporation, discussed earlier in the section and strongly confirmed by my 1991 study. Thirty-nine of the 126 pupils did not recognise the evaporation dimension of the evaporation photograph, believing instead that the water would disappear or soak in its entirety into the soil, the plant or the saucer. Of course, most of the other pupils anticipated some soaking in: what differentiated the thirty-nine is that they did not also anticipate evaporation. Out of the thirty-nine, thirty-seven also denied that changing the saucer was relevant to how quickly the water would go, leading to a highly significant association between denial and awareness of evaporation ($x^2 = 37.05$, df = 1, p<0.001).

Consistent with the literature, the pupils who were unaware of evaporation were concentrated at the younger age levels, with twenty-two for instance in the 6- to 7-year-old group, a concentration that led to a statistically significant age effect ($x^2 = 42.45$, df = 4, p<0.001). This undoubtedly contributed to the fact that with the saucers scene denials of the saucers' relevance were also concentrated at the younger age levels ($x^2 = 23.79$, df = 4, p<0.001). However, it was probably not the only influential factor. For one thing denial was more frequent than lack of awareness of evaporation involving, as we have seen, sixty-seven pupils as opposed to thirty-nine. For another, there was also a statistically significant association between age and denial with the wraps scene ($x^2 = 23.87$, df = 4, p<0.001). It was only the candles scene that did not produce this picture.

When denials were not forthcoming, the pupils readily proposed variables to justify their decisions, and as with the pans and forks scenes discussed in the previous chapter the variables were quite heterogeneous. Thirty-one different variables were proposed for the saucers scene, twenty-eight for the wraps scene, and thirty-nine for the candles. Like the pans and forks scenes, there was a tendency for the most frequently used variables to be scientifically relevant. Yet within this, there were also subtle points of contrast. In the first place, the numbers of pupils using the most and least popular variables were not so markedly different as they were with the pans and the forks scenes (where the most popular variables were used by over thirty pupils and the least popular by one). Although the least popular variables continued to be used by one pupil only with the saucers, wraps and candles scenes, the most popular variable was used by nineteen pupils with the saucers scene, by twenty-one pupils with the wraps scene and by sixteen pupils with the candles scene. Moreover in addition to this, although the single most popular variable was always relevant, irrelevant variables were never far behind. For instance, the number of pupils believing, in complete reversal of science wisdom, that a wide saucer inhibits evaporation was second only to the number subscribing to non-metalness.

The picture is, then, of slightly less of a skew to relevancy than was observed with pans and forks, a picture that was confirmed with reference to the total mentions of variables across the sample. With both the saucers scene and the candles, 45 per cent of the mentions were to relevant variables, and with the wraps scene, the figure was 36 per cent. It will be recalled from the previous chapter that the equivalent percentages for the pans and forks scenes were considerably higher. This said, the mentions of relevant and irrelevant variables were not scattered evenly across the sample for, as Table 4.1 shows, there were age differences in the use of both relevant and irrelevant variables, differences which proved highly consistent across the scenes.

As Table 4.1 shows, there was a tendency for relevant variables to increase in usage with age, a tendency which proved statistically significant with two scenes and nearly significant with the third (F for saucers = 2.51, df = 4,121, p<0.05; F for wraps = 2.35, df = 4,121, p = 0.06; F for candles = 4.87, df = 4,121, p<0.01). Follow-up Scheffé tests showed that the oldest pupils stood apart by using a relatively high number of relevant variables; the youngest pupils stood apart by using a relatively low number. With irrelevant variables, there were further statistically significant age effects (F for saucers = 4.51, df = 4,121, p<0.01; F for wraps = 5.79, df = 4,121, p<0.001; F for candles = 2.85, df = 4,121, p<0.05). However, this time it was the middle age group that the follow-up Scheffés showed to stand apart, for in all three cases the age distribution formed an inverted U shape.

Within the cross-scene consistency, there is one small point of contrast. This is the relatively low number of variables, both relevant and irrelevant, produced for the saucers scene by the youngest age group. The reason for this should be obvious from what has preceded. Pupils who deny that the kind of saucer makes a difference are unlikely to identify saucer variables that facilitate and/or inhibit the loss of water, and denials of this type were

Table 4.1 Mean number of relevant and irrelevant variables as a function of age: phase change (Howe, 1991)

	6–7- *year-olds*	8–9- *year-olds*	10–11- *year-olds*	12–13- *year-olds*	14–15- *year-olds*
Saucers					
Relevant	0.04	0.28	0.21	0.21	0.48
Irrelevant	0.12	0.56	0.75	0.17	0.37
	6–7- *year-olds*	8–9- *year-olds*	10–11- *year-olds*	12–13- *year-olds*	14–15- *year-olds*
Wraps					
Relevant	0.19	0.28	0.17	0.33	0.56
Irrelevant	0.35	0.72	1.25	0.75	0.82
	6–7- *year-olds*	8–9- *year-olds*	10–11- *year-olds*	12–13- *year-olds*	14–15- *year-olds*
Candles					
Relevant	0.12	0.56	0.46	0.33	0.78
Irrelevant	0.65	0.80	0.79	0.42	0.33

both relatively frequent with the saucers scene and clustered at the youngest age levels. Although obvious the point is worth highlighting because the 'denying' pupils included the subsample that was unaware of evaporation, and this subsample produced no variables whatsoever either relevant or irrelevant. Thus, as regards the effects of evaporation awareness upon variable selection (at least in the context of interest) we can conclude by saying that it is a question of some variables rather than none, and not of one set of variables rather than another.

One point that seems to follow from the above is that when we try to explain why the variables selected for phase and temperature change differ in certain respects we should not be calling on the 'distorting' influence of children who are unaware that phases change. This is important for we now have two points of contrast between the present results and what we previously observed for temperature change. The first is, of course, the less extreme demarcation in terms of popularity between relevant and irrelevant variables in the case of phase changes. The second stems from the fact that with phase changes, irrelevant variables show a curvilinear relation with age. It will be recalled from the previous chapter that with temperature change the use of irrelevant variables did not change with age. There might at first sight seem to be a third point of contrast, the fact that with phase changes there were apparently two periods when relevant variables improved, when with temperature change there was only one. However, from the results which have already been presented, the first period of change (between 7 and 8 years) represents a movement from no awareness of variables in this context to an awareness that variables may be operating. It is only the second peiod of change that reflects shifts in the content of beliefs. Thus, if there is a contrast with temperature change here, it is only in the fact that the shift took place somewhat later, between 13 and 14 years as opposed to 10 and 11.

PHASE, TEMPERATURE AND THEORETICAL KNOWLEDGE

The preceding section identified differences between the variables which children call upon for changes of temperature and the variables which they call upon for changes of phase. Yet these differences nevertheless lay within a common framework of age-related improvement of a slow and far from steady kind. When age-related improvement of this nature was identified in the previous chapter for temperature change, it was found to be paralleled by, and arguably dependent upon, an equivalent improvement in understanding of mechanisms. Does the same apply to changes of phase? In other words, can the explanation of the variable growth documented in the preceding section lie with causal mechanisms? The present section will attempt to find out, by exploring the mechanisms which are believed by children to underpin changes of phase and the relevance of these mechanisms to variable selection. Although the evidence is less extensive than that available to the previous chapter, it none the less points in exactly the same direction. Thus, the conclusion will be drawn that children's notions of phase change, like their notions of temperature change, are theoretically structured from a very young age. Faced with this conclusion, the section will end by returning to the theory- and action-based approaches which prompted the book, attempting a preliminary assessment of their standing in the light of what has emerged.

Generative mechanisms for changes of phase

The first step is of course to establish the mechanisms which children treat as causal when they contemplate everyday changes of phase. Do these mechanisms, like the ones detected with changes of temperature, also involve the flowing in and out of two fluids heat and cold? Driver (1985) interprets the results of Osborne and Cosgrove (1983) in a fashion that suggests the situation may be more complex. According to Driver, some of Osborne and Cosgrove's 8- to 17-year-olds did not appreciate the basic scientific truth that although outward appearances alter during phase changes the inherent substance remains the same. On the contrary, they treated phase changes as akin to chemical reactions suggesting that for them the causal mechanisms involve something additional to heat and/or cold. Specifically, just as ash implies coal plus oxygen in the context of heat, so ice may be thought to imply water plus 'X' in the context of cold.

It is possible that a small number of children do think like this, but it is unlikely that the perspective is widespread. In the first place, it does not look as if many children regard changes of phase as changes of substance. In my 1991 study the questions relating to the candles scene were prefaced with 'What will happen to the wax as the wick burns down? Are (melted or non-melted candles) made of the same stuff, or are they made of something different?' Only 26 per cent of the sample responded 'made of something different', with this 26 per cent being concentrated in the youngest two age groups ($\chi^2 = 13.62$, df = 4, p<0.01). Interestingly, equivalent enquiries about scenes depicting bona fide changes of substance elicited equivalent responses. The scenes in question will not be discussed in detail during this book, but one showed the burning of charcoal and the other the rusting of nails. With the charcoal scene, only 38 per cent of the sample said that charcoal and ash are made of something different. With the nails scene, only 35 per cent said this about rusted and non-rusted nails. The response pattern with the charcoal scene was unrelated to age, though with the nails scene age was implicated curvilinearly ($\chi^2 = 11.18$, df = 4, p<0.05). At the youngest and oldest age levels, there was an exactly fifty–fifty split between the pupils saying 'same stuff' or different. In between, the vast majority said 'same'.

In addition, evidence from the candles scene, the charcoal and the nails indicates that very few children see the changes in the 'transformational' sense implied by a chemical reaction. The pupils were asked to explain what happens to wax as it melts, to charcoal as it turns to ash, and to nails as they rust. Only nine gave transformational responses to the candles scene and only fifteen did this to the charcoal and the nails. Transformational responses ranged from the adequate, for example 'The oxygen from the air changes the iron into iron oxide', to the fanciful, for example 'Heat is breaking up the charcoal and putting sulphur dioxide in'. The remaining children either redescribed the event, for example 'The wax gets runnier and then gets harder' or referred to disintegration, for example 'The rain makes the nail get broken down' and 'As it gets burned, it disintegrates into smaller bits which are inside the charcoal.' This pattern of responses is entirely consistent with other research, for in a review of five key studies with pupils up to seventeen years of age, Andersson (1986) concludes that at best only one pupil in five thought in terms of chemical reactions.

The point is then that when matter is transformed in appearance, very few children think that: (a) it has changed in constitution; and (b) it has done more than been broken down or put together. This is the case when the transformation involves a chemical reaction, with constitution being changed and breaking down/putting together being way off beam. It is also the case when the transformation does not involve a chemical reaction, and the childish analysis may not be too far from the truth. With our focus on changes of phase, it is the latter that is of concern to us here. Keeping it in mind, we can assume that most children see, say, melting as the breaking down of solids and freezing as the combining of liquids. Do they, however, see heat and/or cold as playing a crucial role? Moreover do they see the heat/cold as being transferred and/or transmitted? I explored these issues in my 1991 study in relation to the saucers scene and the wraps. As with the pans and forks scenes considered in the previous chapter, opportunities to discuss causal mechanisms (including hot and cold) were provided by follow-up questions to claims about outcomes and claims about variables. These questions were along the lines of 'How does (outcome) happen?' and 'Why is (variable) important?' Such questions were not asked with the candles scene because this was already being used as a context for exploring substance change and chemical reaction. The absence of the questions did not of course prevent some pupils from talking about heat. However, in the circumstances, I felt that formal analysis of their responses might prove misleading. Thus, the following will be limited to what the pupils said in relation to the saucers scene and the wraps.

What the pupils said was relatable to what we established earlier for changes of temperature. Thus, some pupils said that heat would be transferred from the radiator to the saucers and from the sun to the wraps; some said that this would happen and also recognised that transfer implies transmission through some medium; and some seemed ignorant of both transfer and transmission. However within these categories, there were also differences from the two change of temperature scenes represented in the previous chapter. In the first place, ignorance of both transfer and transmission was far more prevalent, manifest in sixty-three pupils with the saucers scene and forty-eight with the wraps. The latter is particularly interesting, for it precludes the simple explanation that ignorance of the mechanism reflected ignorance of the outcome. While, as we have seen, thirty-nine pupils were unaware that water would evaporate from the saucer (and all were in the 'ignorance of transfer/transmission group'), all 126 knew that ice cream exposed to sunshine would eventually melt.

A second point of contrast with the earlier data was the relative rarity of transmissive responses. Nevertheless, such responses occurred and with the wraps scene they increased in frequency with age. Thus, they contributed to the statistically significant age differences in the three categories of response (X^2 for saucers = 28.26, df = 8, p<0.001; X^2 for wraps = 19.30, df = 8, p = 0.01). However, transmissive responses were not the only reason for the age effects. As Table 4.2 shows, 'no transfer' decreased in frequency with age and 'transfer but no transmission' increased. Looking at Table 4.2, it appears that with the saucers scene, the 6- to 7-year-olds were clearly demarcated from the three eldest age groups, and the 8- to 9-year-olds lay somewhere in between. With the wraps scene, three groupings appear: (a) the 6- to 7- and 8- to 9-year-olds; (b) the 10- to 11- and the 12- to 13-year-olds; and (c) the 14- to 15-year-olds.

Looked at in detail, the age-related changes in Table 4.2 do not mirror exactly the equivalent changes over variable selection. There, it will be remembered, the 6- to 7-year-olds and more importantly the 14- to 15-year-olds stood apart as regards relevant variables, while irrelevant variables peaked in the 10 to 11 age group. In view of this, association between mechanisms and

Table 4.2 Mechanisms of heating as a function of age: phase change (Howe, 1991)

Saucers	*6–7- year-olds*	*8–9- year-olds*	*10–11- year-olds*	*12–13- year-olds*	*14–15- year-olds*
No transfer	24	13	9	9	8
Transfer but no transmission	2	12	15	14	18
Transmission	0	0	0	1	1
Wraps	*6–7- year-olds*	*8–9- year-olds*	*10–11- year-olds*	*12–13- year-olds*	*14–15- year-olds*
No transfer	13	15	8	7	5
Transfer but no transmission	10	9	14	14	13
Transmission	3	1	2	3	9

variables cannot be assumed, yet when this possibility was explored correlationally, a pattern emerged that was strongly reminiscent of the pans and forks scenes from the previous chapter. As with pans and forks, there were significant positive correlations between the number of relevant variables and the level of causal mechanisms, the latter being computed via scores of zero to two depending on the Table 4.2 category (r for saucers = +0.31, df = 124, p<0.001; r for wraps (r = +0.36, df = 124, p<0.001). However, there was no association between the number of irrelevant variables and the level of mechanisms.

As with the pans and forks scenes also, there was a strong suggestion that mechanism knowledge was driving variable selection rather than the reverse. As shown in Table 4.3, the percentage of pupils who were aware of transfer but had not generated any relevant variables was 33 per cent for the saucers scene and 40 per cent for the wraps. The percentage who generated relevant variables without apparent awareness of transfer was 6 per cent for the saucers scene and 4 per cent for the wraps. The percentage who understood transmission but mentioned no more than one relevant variable was 1 per cent for the saucers scene and 12 per cent for the wraps. The percentages who mentioned two relevant variables but did not understand transmission were respectively 1 per cent and 0 per cent. When analysed via the Critical Ratio Test, three of the four distributions produced z values which were both positive and greater than 1.96, indicating that variables were significantly more likely to wait on mechanisms than the reverse. The values related to awareness and non-awareness of transfer with the saucers scene (z = +5.00, p<0.001), awareness and non-awareness of transfer with the wraps scene (z = +6.06, p<0.001) and awareness and non-awareness of transmission with the wraps scene (z = +3.87, p<0.001). The only non-significant result related to awareness and non-awareness of transmission with the saucers scene where z was zero.

With the pans and forks scenes it was instructive to repeat the kind of contingency analysis that Table 4.3 has been subjected to for first the two

Table 4.3 Relation between mechanisms and relevant variables: phase change
(Howe, 1991)

Saucers	0 Variable	1 Variable	2 Variables
No transfer	56	6	1
Transfer without transmission	42	19	0
Transmission	0	1	1
Wraps	*0 Variable*	*1 Variable*	*2 Variables*
No transfer	43	5	0
Transfer without transmission	41	19	0
Transmission	9	6	3

youngest age groups taken separately and second the three eldest. However, this was only warranted because there was homogeneity within these group- ings and difference between them across both variable selection and causal mechanisms. As noted above, the age trends with regard to variable selection and causal mechanisms were less consistent for the saucers and wraps scenes than they were for the pans and forks. Thus, contingency analysis for partic- ular age-defined groupings is harder to justify. Nevertheless, it is clear from Table 4.3 that so few pupils in total passed the variable thresholds without the requisite mechanism knowledge that breakdown by age could not bear substantially on the overall picture.

Parallels between phase and temperature change

The differences just noted between the pans and forks age trends on the one hand and the saucers and wraps on the other epitomise what must have emerged from part II as a whole. This is a crisp set of findings relating to changes of temperature which are mirrored in a fuzzier form with changes of phase. Take, for example, variable selection. It is not simply that in my 1991 study the age trends for changes of temperature were immediately and manifestly mappable onto the age trends for mechanism knowledge. It was also that with changes of temperature a limited set of variables stood apart in terms of frequency of usage, and all members of the set were scientifically relevant. With changes of phase, the variables were not so clearly differenti- ated in terms of frequency of usage, and scientifically irrelevant variables were identifiable amongst the most popular. Indeed, the frequency of irrele- vancies was sufficient to highlight an intriguing inverted U distribution as a function of age. Correspondingly for causal mechanisms, my 1991 study isolated three lines of thought for changes of temperature, and the age distribution across the pertinent scenes was more or less constant. The same three lines cropped up for changes of phase, but their usage was different.

There was less consistency across scenes and the overall level of performance was lower. Pupils who showed no awareness of heat transfer with the phase change scenes must have mentioned transfer and perhaps even transmission when it came to changes of temperature.

Without doubt, one factor contributing to the pattern of results was the unorthodox beliefs that young children hold about liquid/gas changes. As we have seen, there is a substantial literature attesting to ignorance of evaporation and condensation in the early school years. My 1991 study confirmed this, and showed moreover what a dampening effect this had on variables and mechanisms. The thirty-nine pupils who interpreted the saucers scene in non-evaporation terms failed to mention any variables whatsoever and fell without exception into the 'no awareness of transfer' camp by virtue of causal mechanisms. However, while unorthodoxies over liquid/gas changes were undoubtedly important, they cannot have been the only significant factor. The wraps scene which featured in my 1991 study related to a liquid/solid change not a liquid/gas, and all pupils anticipated this change perfectly. Nevertheless, the data which the wraps scene generated were more closely aligned with the saucers scene than with the scenes pertaining to temperature change.

Seeking a more complete explanation of why the phase change results were fuzzier, I was reminded of the 'prototype' theory of categorisation (for example, Mervis and Rosch, 1981; Rosch and Mervis, 1975). According to this theory, categories are organised around central exemplars which provide reference points for ascertaining category membership. Thus, robins may be central to the category of 'birds', and it is by virtue of sharing features with robins that 'peripheral' exemplars like ostriches and penguins are regarded as birds. Central exemplars are usually processed in a more definite fashion than peripheral, and developmentally speaking are often prior. Perhaps, then, scenes involving changes of temperature are more central than scenes involving changes of phase, within a category that includes them both.

Accepting this as reasonable, the question is 'which category are we talking about?' This brings us close to one of the book's core issues, for the pointers are to a category defined by causal mechanisms. That is to say 'events involving heating up/cooling down', 'events involving heat/cold transfer' or 'events involving heat/cold transmission', depending in part on the age of the child. My 1991 research revealed significant positive correlations between causal mechanisms and relevant variables. These correlations were apparent for every scene studied, and in all cases they were more likely to reflect the dependency of variables on mechanisms than the reverse. The Howe *et al.* (1995a) study also produced significant positive correlations between mechanisms and variables. However, beside this, it yielded information on the process of knowledge change, and that information signalled a driving role for mechanisms. Obviously, there is a pressing need for additional research. Nevertheless, it is significant that the existing research is entirely consistent in what it suggests and moreover that it offers no qualification as a function of temperature vs. phase nor of children's age. In the 1991 study, mechanisms were as likely to be generative for the 6 to 7 age group as they were for the 14 to 15. What changed with age was the form that the mechanisms took, and hence the variables that they were associated with. Their generativity, as far as can be discerned from the data, remained entirely constant.

The absence of age trends over generativity is of course an extremely important finding. It suggests theoretical structure from the early years of schooling when, as emphasised repeatedly, the Piagetian model denies this until sometime after 10. Thus, Piagetian theorising is challenged by the present results, and the fact that the challenge occurs in the context of heat transfer is in reality highly ironic. Theoretical knowledge of heating and cooling was identified by Piaget (1974) as particularly inaccessible to children. It was unfavourably compared with object motion, as lacking in transparency to everyday awareness. We shall be considering object motion in part III, and when we do Piaget's views might usefully be borne in mind. For now, the only issue is what we do about heat transfer given that theorising manifestly was accessible around the age of 6, and hence on this point Piaget was wrong.

One possibility is to shift to Piaget's action-based counterpart, Vygotsky. Vygotsky allowed for theorising as early as 6, and with a topic like heat transfer could have been said to predict this. The trouble is that while predicting early theorising Vygotsky runs into difficulties with other aspects of the present results. There would be problems for Vygotsky in the data relating to temperature change which indicated mechanisms and variables becoming more orthodox at 11 or 12, after long periods of limited progress. Leaps forward after delays were also observed with changes of phase and, as befits their arguably peripheral status in the prototype, these were less marked and chronologically even later. This too would prove problematic, for the Vygotskyan prediction is improvement throughout the school years. A leap forward might be accommodated if, for example, formal teaching started and pupils were exposed to new conversational practices. However, this possibility has no bearing on the present results. The 10- to 11-year-olds grouped with the older children on most analyses involving my 1991 data. However, as noted in the appendix, this age group had not received instruction in any aspect of physics. Indeed, the 12- to 13-year-olds who had certainly begun physics had not reached the transfer of heat. Quite apart from this though, the Vygotsky model would also predict improvement prior to 10, and this was not typically observed.

Nevertheless, despite the problems we should probably need a model which continued some aspects of the Vygotskyan perspective if we wanted to sustain an action-based perspective in the face of the present results. In particular, we should most likely need to call upon co-ordinations between schemes and their symbolisation in language. Such co-ordinations would seem to be the only possible sources of early theoretical structure within an action-based framework. However if we accept scheme–language co-ordinations, we should also have to accept that, driven by the desire for communicative competence, children will strive for mature conceptions of mechanisms and the variables they generate. These conceptions will indeed be part-and-parcel of the same linguistic representations which led to theoretical structure in the first place. Having accepted this, we should therefore have to deal with the absence of progress prior to 11 by imagining something intervening to prevent the mature conceptions from becoming stable aspects of cognitive structure until some years later. Given the age trends, this 'something' would probably operate to undermine the *protective* function of language that was outlined in part I. After all, without protection from language children would, on action-based models and as spelled out by Inhelder and Piaget (1958), fall into the vicious circle of aimlessly shifting beliefs, a circle which is only broken when around 11 entity-action differentiation is sufficient to allow a possibilistic perspective on mechanisms and variables.

The key ingredient from the action-based perspective is, then, a wedge which is driven between language and the conceptual knowledge of children. If we wanted to salvage the perspective, we should need to formulate convincing proposals as to what the wedge was like. However, do we want to salvage the perspective? Should we not jettison action-based theorising, and shift instead to the theory-based approach? Perhaps, but as noted already the theory-based approach would also have difficulties with the present chapter's results. It is true that the approach posits theoretical underpinnings to all conceptual distinctions. Thus, insofar as children become aware of heating up and cooling down by the early years of schooling, possession of relevant theories is predicted by that age. However, once theories exist, their improvement should be regarded as functional. Thus, steady, age-related improvement in mechanisms (and as a consequence variables) would be expected just as it would be given the Vygotskyan stance. Here again then, the substantial progress around 11 years of age but not before would seem to cause problems, and this time there is no obvious escape route in the form of a potential wedge. The evidence is currently too finely balanced to call, but we might in view of this apparent inflexibility move to part III feeling that all may not be well for theory-based thinking.

Part III Propelled motion

5 Encapsulated knowledge of horizontal motion

Our basic question is how do school-age children think about variables and mechanisms in areas that are relevant to physics. We wish to know which variables and mechanisms they recognise, and how they organise these variables and mechanisms in their knowledge structures. In part II we asked the question in relation to the transfer of heat. We found that children recognise a number of variables, many of which are consistent with received science orthodoxy. However, while this was true at all levels of schooling, there was evidence of enhanced understanding with age and in particular of substantial progress around or after 11 or 12. A similar picture emerged with mechanisms. The majority of school-age children appreciate that heat is transferred, but recognition of the transmissive nature of transfer is rare before the teenage years. When we shifted attention to knowledge organisation, it became apparent that the parallel age trends for variables and mechanisms were probably not accidental. There was a close association between progress over variables and progress over mechanisms which was suggestive of contingency. In particular there were signs that progress over variables is dependent upon progress over mechanisms, indicating a generative role for the latter in relation to the former.

Without doubt, the evidence for the generativity of mechanisms was far from watertight. Nevertheless, it was as strong for the youngest age groups as it was for the oldest, indicating as part II emphasised, that knowledge of heat transfer may be theoretically organised from the early years of schooling. As such, the evidence has immediate and negative implications for the Piagetian approach, for there theoretical structure would be denied until at least 11 or 12. However, as was also brought out in the previous chapter, the evidence has equally negative implications for the Vygotskyan approach. Although this approach could in principle accommodate early theorising, it would predict a steady improvement in mechanism and variable knowledge from around the age of 6. Thus, it would be stymied by the dramatic progress in both forms of knowledge from 11 onwards, after zero progress for five or so years. Equally bothered though would be the theory-based approaches which seemed in part I to offer so much to educational practice. These approaches would also be comfortable with early theorising;

indeed this is a central tenet of their thesis. However, they too would anticipate steady improvement in mechanism and variable knowledge from the
very beginning.

Ending with the acknowledgement that none of its guiding models could
fully explain the results, part II sketched the options for developing the
models into more adequate forms. Very tentatively, it indicated that the
action-based framework which both Piaget and Vygotsky reflect may have
more flexibility than its theory-based alternative, and thus may be the route
to follow. An outline was provided of how the Vygotskyan approach might
be developed to leave children more open to the processing difficulties which
Piaget described. This said, the previous chapter was tentative, for its
endorsement of the action-based perspective was founded solely on the
potential for modification. Summing up what the evidence has shown, the
fair verdict would be consistency and inconsistency with both perspectives
to equivalent degrees.

Clearly then, there is a need for further, more incisive evidence, but where
could it come from? By way of an answer, it must not be forgotten that in
taking heat transfer as its topic, part II was focusing on an area which, as
was pointed out in part I, provides some of the strongest evidence for theoretically structured knowledge in the post-childhood years (and specifically
in students who have reached an age to be embarking on physics teaching).
There are other topic areas where the evidence relating to theoretical structure in older students is either implicit or controversial. Of course if any of
these areas turned out on closer investigation to lack theoretical structure at
the point of formal instruction, this would in itself deliver a killing blow to
the theory-based perspective. However, even if this did not happen, the areas
in question are obviously ones where theoretical structure is relatively
opaque. In which case, there should be an equivalent opacity about linguistic
representation, meaning of course that even on Vygotsky's version of action-
based theorising the acquisition of theoretical structure should be delayed.
Since linguistic representation is not the source of theoretical structure from
the theory-based perspective, the prediction here remains early theorising.

In view of all the above, cross-area comparison would appear highly
desirable, and anticipating this, part I identified propelled motion as a
potentially fruitful contrast with the transfer of heat. As part I made clear,
extensive research has been conducted with contexts where objects which are
propelled into space begin to fall. Based on this research, some scholars have
argued for theoretical structure in students who are embarking on physics
teaching and some have denied this. Thus, unlike heat transfer, there is
dissension and debate regarding everyday physics in its full-fledged form.
What does this imply for the thinking of children? In the present chapter
and the next one, we shall attempt to find out. Paralleling our discussion of
the transfer of heat, we shall consider the variables that children deem relevant to movement after propulsion, and the contingency of those variables
upon causal mechanisms. To avoid undue complexity, we shall build up

gradually to contexts where objects are propelled into space. Such contexts have a horizontal and a vertical dimension, horizontal as objects move forwards into space and vertical as they dip towards the ground. Much simpler are contexts like curling stones on an ice rink or wooden balls in a bowling alley where the motion is purely horizontal. Thus, these are the contexts with which we shall start in the present chapter. We shall add verticality in its successor.

SPEED AND ITS GOVERNING VARIABLES

Motion in a horizontal direction is something that children experience frequently in their daily lives. Quite apart from the sporting events mentioned above, there are the games that children play as they roll toy cars across the floor or aim computer icons across the screen. There are in addition the journeys that children make, either under their own steam by, say, bicycles or in the cars their parents drive. Children also witness horizontal journeys being undertaken by others as aeroplanes cross the sky or trains pass the embankment. In many cases, these everyday experiences involve propulsion, that is objects are given a push to set them in motion but there is no subsequent force in the travelled direction. Knowing this we can safely assume that propulsion in a horizontal direction is something with which children are familiar. However, what does familiarity mean? Does it mean something which is taken for granted or something which is recognised as variable and analysed accordingly? The screams of children at swing parks, 'Faster, faster' or 'Slow it down', make it clear that there is one dimension, at least, along which motion is analysed, speed. This is interesting for speed is close conceptually to velocity, and velocity is central to all formal analyses of object motion. Thus, with speed, we have a concept which is both meaningful to children and linkable to science, and this seems to recommend it as a focus for our discussion. Accordingly, the questions to be asked in the present section will relate to the speed of horizontally propelled motion. The section will start with the criteria by which children determine speed, and move to the variables which they call upon in relation to speed.

Children's criteria for estimating speed

Within science, the relation between speed and velocity is straightforward. Velocity is speed in a given direction. However, although straightforward, the relation is seldom accessible to children, for an independent concept of velocity is rarely acknowledged. For instance, Jones (1983) found that only one pupil in a sample of thirty 11- to 16-year-olds could define velocity correctly. Thirteen admitted to having no idea what the word meant and the remainder confused it with speed. Thus, what for science is a relation turns out for children to be a fusion at best. Perhaps though, it is a fusion which maintains the sense of speed that is recognised by science. It would seem sensible to discuss this issue before we consider variable selection or causal mechanisms, and as it happens, there is a lot of relevant research. This is due directly to work by Piaget that was published in English translation in 1970 and indirectly to Einstein's elaboration of Newton's Laws.

To explain, one of Newton's many contributions to science was to show that for all practical purposes velocity (v) can be construed as distance (d) per unit time (t), that is $v = d/t$ Thus, for Newton, an adequate conception of velocity depends on co-ordinated knowledge of distance and time. Einstein had no quarrel with this, but nevertheless his work on relativity convinced him that what is adequate for the everyday world is not adequate for the limit cases. In the limit, velocity has to be as basic a concept as time, and thus is inaccurately construed as dependent upon it. Recognising this, Einstein wondered whether velocity and time might be equally separated in the thinking of children, with $v = d/t$ in effect a subsequent construction. In 1928, Einstein raised the matter with Piaget, thereby stimulating the work reported in Piaget (1970).

The substance of Piaget (1970) is a series of studies where pupils aged 5 to 14 were asked to judge the relative speed across specified time periods of pairs of moving objects. The studies varied along several dimensions, including whether or not one object overtook another, whether or not the objects started together or stopped together, and whether or not the full motion was visible rather than partially eclipsed by a tunnel. Based on the judgments that children made (and the way that their judgments were justi-fied), Piaget drew two conclusions that would have been grist for Einstein's mill: (a) even very young children have a secure notion of speed; and (b) this notion is not initially based on distance per unit time. Becoming more specific, Piaget further argued that despite not being based on distance per unit time, young children's notions of speed are none the less systematic. They are founded in particular on overtaking and finishing, for objects which overtook, finished ahead or reached some stopping line first were regarded by many pupils as travelling at a faster speed. This is, of course, less encouraging from the perspective of Einstein for it means that the sepa-ration of speed from time does not reflect latent relativity in the minds of young children. Quite the opposite, it signifies a concept of speed that is not even as good as distance per unit time.

Of course, Piaget's conclusions may be wrong, but I myself think this unlikely. As intimated above, the 1970 book has stimulated a fair amount of follow-up, and my reading of the results is support for Piaget rather than opposition. Typical here is a study by Siegler and Richards (1979) which extended Piaget's analysis of the significance of stopping position. Twelve 5- to 6-year-olds, twelve 8- to 9-year olds, twelve 11- to 12-year-olds and twelve adults observed two trains running on parallel tracks. Starting position, stopping position, starting time and stopping time were all varied to produce problems where stopping position would or would not provide a valid index of speed. Consistent with Piaget, Siegler and Richards found many 5- to 6-year-olds relying on stopping position even when it produced the wrong answer. In particular if one train stopped ahead of another, it was considered to have travelled faster. Moreover, it was not until 11 or 12 that distance per unit time was consistently used.

In like vein, Mori *et al.* (1976) report a study with sixty-one Japanese children aged nearly 5 to nearly 6. In this study, the children were interviewed about six problems involving toy cars where reliance on stopping position would almost always have yielded the wrong answer. Prior to formal teaching, only two children answered more than two problems correctly. There is some ambiguity about these results in that the children were only quizzed about reasons for their answers when they responded correctly. Nevertheless, even amongst correct responders there was clear evidence for reliance on stopping position. Interestingly, although formal teaching improved the level of accuracy with 60 per cent now giving more than two correct answers, it did nothing for problem-solving strategy. On the contrary, the main reason for improved accuracy was enhanced reliance on perceptions of which cars 'went whizzingly' (Mori *et al.*, 1976: 525). By some reckoning, this could be regarded as a regression.

The failure of Mori *et al.*'s teaching intervention to enhance problem-solving strategy seems indicative of considerable resistance to distance per unit time. However, if this is the case, Mori *et al.*'s results appear hard to reconcile with a claim made by Wilkening (1981). This is that the absence of distance per unit time in Piaget's data and those of Siegler and Richards should not be treated as a dearth of knowledge. Rather, it should be interpreted as a deflection from knowledge by extraneous cues such as overtaking and/or stopping position. The implication is that if the distracters were removed (and Mori *et al.*'s intervention should have done this), distance per unit time would readily have shone through. Certainly, Wilkening reports a study which could be read as endorsing the point, for it is consistent (in some respects at least) with children as young as 5 using two of distance, time and speed to infer the third.

Wilkening's study involved cartoon animals of differing 'natural speeds', for example a snail and a deer, fleeing every time a fierce-looking dog began to bark and stopping once the barking had ceased. To check the inference of distance from speed and time, the task was to predict how far each animal would have run after barking of varying durations. To check the inference of time from distance and speed, it was to predict how long the dog had barked given the distance travelled by each animal. Finally, to check the inference of speed from distance and time, it was to predict which animals would have run specified distances after specified bouts of barking. Forty-five 5-year-olds, forty-five 10-year-olds and forty-five adults participated in the study, with fifteen subjects at each age level performing each task.

Regardless of age level, Wilkening's subjects produced response profiles consistent with the beliefs that $d = v/t$ and $t = d/v$. It is on these profiles that Wilkening rested his case. Nevertheless, there was no evidence at any age level for $v = d/t$. Rather, the response profiles of the 5-year-olds were consistent with $v = d$, reminiscent surely of Piaget's stopping position. The response profiles of the 10-year-olds and the adults were consistent with $v = d-t$. Thus with a task which attempted to put speed into focus, there were no signs of a Newtonian perspective. Since speed in focus is precisely the concern of this chapter, this could be construed as a result of some significance. The point is worth taking but to rest matters there would, I feel, miss something crucial. This is that by using the natural characteristics of animals to establish speed, Wilkening may have done precisely what he accused his predecessors of doing, and deflected his subjects' attention away from speed. In particular, attention may instead have been on something akin to 'power', and indeed by assigning differential power values to the animals, say P1 to a snail and P5 to a deer, response profiles akin to $d = vt$ and $t = d/v$ could have been obtained, without knowledge of speed.

Certainly, an Australian study by Cross and Mehegan (1988) suggests that something must be amiss with Wilkening's methodology, for poorer performances were observed with apparently simpler procedures. In Cross and Mehegan's study, which involved 175 children aged 4 to nearly 10, information was supplied about the differential speeds of two cars over periods of time. Using this information, the children were asked to choose the routes (from options on a layout) that the cars would have to follow to reach a goal at the same time. They were also asked whether the arrival times would be the same if the routes were equivalent. Thus, distance was to be predicted given time and unmistakably speed, and time was to be predicted given distance and equally unmistakably speed. Cross and Mehegan's procedure was simpler than Wilkening's in at least two respects. Two cars were used whereas Wilkening's tests involved three animals. Moreover, when distance was given, it was the same for each car and likewise for time. All Wilkening's given variables differed. Nevertheless, despite the procedural simplicity, Cross and Mehegan's studies intimate major conceptual weaknesses in the children's thinking. Success on distance prediction varied from 4 per cent of the under-fives to 56 per cent of the over-nines. The corresponding figures for time prediction varied from 32 per cent to 94 per cent.

Thus, even in the rarefied context of controlled developmental research, children have real difficulties with the notion of distance per unit time. How much more problematic the concept must be when children draw inferences in the real world. There, it will not only be a question of dealing with overtaking and stopping position instead of being protected from them. Rather, there will be a need also to collate from events that are separated in time and space, thinking how an old tennis ball worked when it was new and how the bounce on a grass court differs from the bounce on a clay. Additionally, there will be the problem of taking frame of reference into account, for motion viewed from a moving vehicle will look different from motion viewed from a stationary position.

The implications of such variables came home to me thanks to a study that I worked on a few years ago (Howe *et al.* 1992a). This study involved computer simulations where two trains moved across a screen. Sometimes the trains started together; at other times one train moved after the other. Sometimes, both trains moved from left to right; at other times one moved from right to left. Finally, sometimes the motion was to be viewed from a stationary position; at other times it was to be viewed from the perspective of a third moving train. The task in all cases was to identify the single point at which the speeds of the trains were identical (given that one train was moving at constant velocity while the other was accelerating or decelerating). The subjects in this study were not school children but rather undergraduate students. Many were studying physics at university and even those who were not had often obtained a physics 'Higher'.[2] Despite this, accuracy of problem solving was low, and interview responses indicated that few subjects were referring to distance/time data. This was in spite of the fact that graphical representations of distance per unit time and time per unit distance could be called up on-screen. In quantitative terms, consistent reference to distance/time data would have obtained subjects a mean score of at least 4.00 on the scale shown in Table 5.1. In fact the mean score was 2.52.

Variables relevant to horizontal motion

The conclusion to be drawn from the preceding paragraphs is that, although speed is a meaningful concept to children, they do not have reliable procedures for ascertaining it in the real world. Thus, when they size up a situation and say 'That went relatively quickly', they may well be in error. Does this mean then that their database for deciding what makes things go quickly and what makes them go slowly must of necessity prove shaky? If children make such decisions with reference to attributed speed of travel, the answer would seem to be 'Yes'. However if they behave like professional scientists, this is not how they will be deciding. Rather they will be referring

Table 5.1 Strategies for comparing velocities (Howe *et al.*, 1992a)

0	represented failure to articulate any strategy
1	represented reliance on visual judgment alone
2	represented an erroneous conceptual basis (e.g. 'the trains must have been going at the same speed when they passed each other')
3	represented a partially correct conceptual basis (e.g. reference to 'the gaps being the same' on the time/distance diagrams)
4	represented a correct but non-relativistic conceptual basis (i.e. explicit statement that two objects must be at the same speed if they take the same time to travel the same distance, but no recognition that distance and speed are measured with respect to some framework)
5	represented a correct and relativistic conceptual basis.

to speed change, that is acceleration or deceleration. Attributed speed of travel will be beside the point. We do not currently know how children will behave. However, let us imagine for the moment that they do resemble scientists. Let us imagine also that they do not merely use the information that scientists refer to but also process the information in an equivalent fashion. Their inferences about variables should be both rich in range and complex in operation.

To see why, consider first a world which is much simpler than the present one, a world where object movement takes place in a vacuum. In such a world, the sole force (F) when objects are propelled would be the push that objects receive to set them in motion. Let's call this force P, and note (as we did early in part I) that it ceases to operate when propelled objects are moving independently. The importance of this stems from a celebrated discovery of Newton's (expressed as his Second Law) that acceleration (a) is directly related to force and indirectly related to how heavy the object is (m), that is $a = F/m$. It follows that in vacuo acceleration at the start of independent motion will be P/m; acceleration thereafter will be $O/m = 0$. The significance for object velocity (and, of course, speed) comes from the fact that velocity at the end of some time period (v) is known to be equivalent to the initial velocity (v_o) plus the product of the acceleration and the time elapsed (t), that is $v = v_o + at$.

Thinking once more of curling and bowling and considering the period from the start of a curler's/bowler's build up to the moment of independent motion, v_o would be zero since initially all is at rest and at would be Pt/m. Thus, velocity at the start of motion would be Pt/m. Considering by contrast the period from the start of independent motion to some arbitrary time afterwards, v_o would now as just established be Pt/m but at would be $Ot = O$. Thus, velocity at any time after the start of independent motion would also be Pt/m. Newton saw this fact as highly significant for he enshrined it in his First Law: if a body on which no net force acts is at rest, it will remain at rest; if it is moving, it will continue to do so with constant velocity.

The central points for us are that in a vacuum: (a) independently moving objects will sustain their initial velocities; and (b) these initial velocities will be directly related to the force of propulsion and indirectly related to how heavy the objects are. Put more simply, given equal thwacks, light objects will always be travelling faster than heavy ones. This established, let us now remove the 'in vacuo' assumption and consider the qualifications required by placing objects in real-world contexts. In the real world, objects moving independently in a horizontal direction are subject to two kinds of opposing forces: (a) frictional forces from the surfaces that objects touch; and (b) drag forces from the fluids that objects pass through. Friction is known to be the product of m, the force of gravity (g), and a coefficient which decreases as surface slipperiness increases (μ), that is μmg. Drag is known to be half the product of the fluid's density (p), the object's surface area (A), the square of the object's current velocity (v^2) and a coefficient which decreases as object smoothness and streamlinedness increase (C), that is $CpAv^2/2$.

Substituting the above formulae into the one derived above from Newton's Second Law, we have $a = -\mu mg/m - CpAv^2/2m$. (Friction and drag are expressed as negatives because they operate in the opposite direction from motion.) Simplifying, we have $-a = \mu g + CpAv^2/2m$, with 'negative acceleration' ($-a$) being equivalent of course to deceleration or slowing down. Thus in the presence of friction and drag, independently moving objects will not travel at constant velocity but rather will gradually come to rest. Thanks to friction, the rate at which they will do this depends on gravity and surface slippiness. The rougher the surface, the faster the rate will be. Thanks to drag, the rate at which objects will achieve rest depends on fluid density, object surface area, object velocity, object shape, object smoothness and object heaviness. The denser the fluid, the larger the surface, the faster the velocity, the less streamlined the shape, the rougher the texture and the lighter the objects, the faster will be the rate to rest.

This, then, is where children would get to if they behaved like professional scientists. There would, indeed, be a rich array of variables. Moreover, if we look carefully at what has been claimed for one variable, object heaviness,[3] we can see why the array was earlier deemed complex. The point about object heaviness is that it does not operate in a unidirectional fashion. All other things being equal, light objects will be propelled into motion with greater velocities than heavy ones. However, due to drag, light objects will slow down faster than heavy ones. Thus, in some circumstances, it would be legitimate to say that lightness facilitates movement. In others, it would be legitimate to mention heaviness. On the face of it, it seems highly unlikely that children will appreciate this paradox unless they are privy to the received wisdoms of science. It is, for example, surely inaccessible to direct observation. However, are children privy? To begin to find out, let us look at the variables they use in practice. Let us see whether they mesh with the list presented in the previous paragraph and, if they do, whether object heaviness is used in a half-way adequate fashion.

The first study to give us any inkling of answers lies amongst a set reported by Inhelder and Piaget (1958). The focus of the study in question was a piece of apparatus whereby balls could be launched via a spring device to roll along a horizontal track. Balls of differing weight and size were provided, and the task was to ascertain the variables relevant to distance travelled. Inhelder and Piaget report responses to the task by pupils aged 5 to nearly 15 (although they omit formal details of the age range covered). Responses involved predictions of the distances that the balls would roll and accounts of why these distances were predicted. Predictions/accounts were used to categorise each pupil into one of three stages of reasoning, the first being characterised by unwitting contradiction. Typical was the 6-year-old who argued that a little wooden ball 'won't go very far because it's small'; a large wooden ball 'can't go very far because it's big'; and 'the two big ones will go less far because they're big . . . the three little ones won't go as far as the big ones' (Inhelder and Piaget, 1958: 126).

By the second of Inhelder and Piaget's stages, contradiction was believed to be recognised. However, 'in spite of the effort to eliminate this, a residue of contradiction is left from the fact that the heavy balls have a greater force when in motion but are less easily set in motion, whereas the light ones have less force but are more easily launched' (Inhelder and Piaget, 1958: 167). In other words, the paradoxical phenomenon of object heaviness which was outlined above is a major stumbling block for the second stage pupil. All is believed to be sorted out by the third stage, which is exemplified in Inhelder and Piaget's text by pupils aged 12 to 15. Thus typical here was the view of one pupil aged 14-and-a-half that 'A heavier ball takes off less easily but goes further because it has force in itself' (Inhelder and Piaget, 1958: 129). This pupil, like many others of a similar age, equally argued 'No, the bigger they are, the stronger the air resistance' suggesting that the dynamics of size were also worked out.

Although Inhelder and Piaget's conclusions are intriguing, they do raise some questions. We are not told the proportion of 12- to 15-year-olds attaining the third stage. However, it is clearly more than a handful and given the complexity of the issues, this seems, on the face of it, surprising. Thus, the first question must be whether the apparent precocity of Inhelder and Piaget's sample is replicable elsewhere. In addition though, there are issues relating to the apparent centrality of size and weight at all three stages. This is also precocity of a sort for size and weight are relevant dimensions, even if the use of the dimensions was not free from contradiction in the first and second stages. However, are size and weight the only variables that children consider? If so, this would be out of line with what part II established for the transfer of heat. There, it will be remembered, the number of variables called upon could reach twenty or thirty. There may be bona fide differences between the topic areas. Nevertheless, it has to be recognised that Inhelder and Piaget's materials included balls selected to vary in size and weight. It could be, therefore, that the choice of materials rendered size and weight unusually prominent. This intimates a second question, namely would a less focusing set of materials have broadened the variables that children refer to?

Recognising the questions, I included two items in the Howe (1991) study aimed at providing answers. The first related to a photograph where a large green ball was being rolled across paving stones as in hopscotch, with a tennis ball, table tennis ball, golf ball and bowls wood in the immediate vicinity. The large green ball was identified to the pupils as being made of solid plastic; the other balls were identified as normal exemplars of their genre. The second item related to a photograph where the bowls wood was being 'curled' across an ice rink, with the other four balls from the hopscotch photograph in the immediate vicinity. With both scenes the pupils were asked questions along the lines of 'Do you think that, no matter how hard you push, some kinds of balls could always be made to roll faster than others? Why? Has the player chosen the ball that could be made to roll the fastest? Why? Which ball could be made to roll the fastest? Why? What about the balls that couldn't be made to go so fast? Would they all go equally slowly or could some be made to go faster than others? Why?' As it happened, nine pupils in the sample of 126 denied that some balls could be made to go faster than others with the hopscotch scene and twenty did this with the curling. In both cases denials were concentrated in the two youngest age groups (χ^2 for hopscotch = 14.45, df = 4, p<0.05; χ^2 for curling = 10.60, df = 4, p<0.05). These age groups, it will be remembered, were the 6- to 7-year-olds and the 8- to 9-year-olds.

Despite the denials, a clear majority at every age level thought that the kind of ball would make a difference, and in all cases they were able to pinpoint one or more variables to justify their views. Across the sample as a whole, twenty-two different variables were generated for the hopscotch scene and twenty-one for the curling. Since the average number of variables per pupil was 1.51 for the hopscotch scene and 1.25 for the curling, it can be inferred that the pupils differed between themselves in which variables they mentioned. In fact with the hopscotch scene, only five variables were used by more than ten pupils. Listed in terms of aids to going fast, these were heaviness (forty-seven pupils), lightness (thirty-nine), smallness (twenty-seven), bigness (nineteen) and bounciness (sixteen). With the curling scene, four variables were used by more than ten pupils. As aids to going fast, they were heaviness (fifty-one pupils), smoothness (thirty-two), lightness (twenty-one) and smallness (twelve). Relating these data back to Inhelder and Piaget (1958) three points emerge: (a) even though my sample included pupils up to 15 years of age, there was no evidence for the co-ordinated use of heaviness *and* lightness which typified Inhelder and Piaget's third stage of reasoning; (b) heaviness *or* lightness were used frequently by the pupils in my sample, as were bigness or smallness, implying in accordance with Inhelder and Piaget that weight and size were significant variables; and (c) although weight and size were significant, they did not account for all the variables: bounciness and smoothness appear amongst the 'consensual' variables listed above and with both scenes there were seventeen further 'idiosyncratic' variables, used in fact by between one and ten pupils.

Because some pupils were citing smallness and smoothness as conducive to going fast, there was clearly some scientific orthodoxy in the thinking of my sample. It will be remembered that drag increases as size and roughness increase, and that the bigger the drag the greater the deceleration. There was however also irrelevancy, as witnessed not just by bigness and bounciness as facilitators of speed but also by many of the idiosyncratic variables. To obtain a feel for the latter, it should be noted that they included softness, dryness, plasticity and holeyness as facilitators of speed. Faced with both relevancy and irrelevancy, it was of interest to ascertain how either or both changed with age. Thus, taking the responses to each scene separately, I counted the number of relevant and irrelevant variables that each pupil mentioned and calculated the means as a function of age. I excluded heaviness and lightness from my count because they cannot unambiguously be regarded as relevant or irrelevant. With the hopscotch scene, neither the use of relevant variables nor the use of irrelevant changed with age (F for relevancy = 0.96, df = 4,121, ns; F for irrelevancy = 1.21; df = 4,121, ns). With the curling scene, there was no change with age in the use of irrelevant variables (F = 0.96, df = 4,121, ns). However, there was a statistically significant increase with age in the use of relevant variables (F = 4.10, df = 4,121, p<0.01). The oldest of the five age groups, the 14- to 15-year-olds, used more relevant variables than any other group, differing significantly on Scheffé tests from the 6- to 7-year-olds, the 8- to 9-year-olds and the 10- to 11-year-olds.

The significant age difference with the curling scene over the use of relevant variables stemmed from the frequent reference to smoothness by the 14- to 15-year-old subjects. Indeed, fourteen of the thirty-two pupils invoking smoothness as conducive to going fast were in the 14 to 15 age group, a concentration that led to a statistically significant age effect for that variable alone (χ^2 = 17.18, df = 4, p<0.01). In this context, it is of interest that although only five pupils referred to smoothness with the hopscotch scene, all but one were in the two oldest age groups. Thus, smoothness was a discovery of the later school years, but so in addition was heaviness. As Table 5.2 shows, there was with both scenes a statistically significant increase with age in the belief that heavy objects move relatively fast. The effect of this was that a majority of the two oldest age groups called upon heaviness. With the younger pupils it was a clear minority. As Table 5.2 also shows, there were some indications of a decrease with age in references to lightness. However, this was not statistically significant. In fact, none of the other variables identified earlier as consensual was used differentially as a function of age.

Table 5.2   Use of lightness and heaviness as a function of age: horizontal motion (Howe, 1991)

	6–7-year-olds	8–9-year-olds	10–11-yea- olds	12–13-year-olds	14–15-year-olds	χ^2 for age (df = 4)
Hopscotch						
Heavy things go fast	5	6	8	12	18	16.88, p<0.01
Light things go fast	9	11	9	5	4	7.14, ns
Curling						
Heavy things go fast	7	6	8	14	15	11.07, p<0.05
Light things go fast	4	8	5	2	3	6.08, ns

Obviously, the picture is complicated by the popularity of heaviness and lightness, variables which it is hard to characterise in terms of relevance. Nevertheless, it shows a few points of contact with what we observed in part II for the transfer of heat. As with heat transfer, children call on a wide range of variables, wider indeed in my 1991 study than what published research implies. These variables are a mixture of the scientifically relevant and the scientifically irrelevant, relevant variables showing stronger evidence than irrelevant of increased usage with age. When we observed this picture in part II, we subsequently discovered that variable selection may well have been dependent upon implicit theories. Does the same apply for propelled motion in a horizontal direction? In other words, do children have ideas regarding the causal mechanisms that determine horizontal motion, and do they bring these to bear on variable selection? These are the issues which we shall be discussing in the section to follow.

FORCE AND HORIZONTAL MOTION

Sufficient physics has already been presented to establish the scientific orthodoxy on the causal mechanisms which apply to horizontal motion. In particular when objects are propelled in a horizontal direction, they will take on a velocity that is a function (in part at least) of how strongly they were pushed. Once in motion, objects propelled in a horizontal direction will be subject to friction and drag, and unlike the more complex motion to be discussed in the next chapter these will be the only forces of relevance. Since friction and drag oppose motion, the consequence will be constant deceleration, meaning that objects propelled in a horizontal direction will begin to slow down the instant they are released and hence that at no point will their velocity exceed the initial level. This orthodoxy should be borne in mind when contemplating the ideas of children about causal mechanisms, for it will provide a useful heuristic for organising the disparate (and by no means internally consistent) material that the present section will attempt to interpret.

Impetus and external push

Even before we consider the mechanisms that children refer to, we can find hints in my 1991 data that they will not be very advanced. Rather than launching immediately into the questions outlined earlier, presentation of the hopscotch and curling scenes was accompanied by 'What will happen to the ball's speed as it rolls across the paving stones/ice rink?' Analysis of the pupils' responses revealed that only nine anticipated immediate deceleration with the hopscotch scene and only five with the curling. The views of the remainder are shown in Table 5.3 from which two points can be gleaned: (a) with the hopscotch scene, the most common view by far was to anticipate slowing down after a period of steady speed; and (b) with the curling scene, a majority of the sample did not anticipate any slowing down while the ball was on the ice.

Table 5.3 Anticipated speed change during horizontal motion (Howe, 1991)

	6–7-year-olds		8–9-year-olds		10–11-year-olds		12–13-year-olds		14–15-year-olds	
	Hopscotch	Curling	Hopscotch	Curling	Hopscotch	Curling	Hopscotch	Curling	Hopscotch	Curling
No comprehension										
Don't know	1	1	0	0	0	0	0	0	0	0
Ball will get faster and faster	4	10	1	12	1	11	2	9	2	5
Ball will move at steady speed	3	3	2	3	3	4	0	3	1	5
Partial comprehension										
Ball will begin to slow down after speeding up	1	4	3	2	1	5	3	6	4	8
Ball will begin to slow down after steady speed	17	8	19	8	18	4	16	4	15	6
Full comprehension										
Ball will begin to slow down immediately	0	0	0	0	1	0	3	2	5	3

Table 5.3 suggests modest improvement with age in level of understanding. To gauge its statistical significance, I conducted ANOVAs on scores obtained by assigning zero for 'no comprehension', one for 'partial comprehension' and two for 'full comprehension'. The outcome was a significant improvement with age for the hopscotch scene (F = 2.91, df = 4,121, p<0.05) but not for the curling (F = 1.32, df = 4,121, ns). Thus, the 15-year-olds were just as likely as the 6-year-olds to believe that balls on ice sustain their speed. To return to the point, it is hard to imagine them holding this view and simultaneously endorsing the received gloss on causal mechanisms.

However, to find out for sure, we need to address mechanisms directly. When we do, we find the literature pointing up a major confusion which mirrors exactly what was reported in part I for horizontal motion in the context of falling. This can be summarised as a strong commitment by children to forces in the direction of motion for the duration of motion, when as we have seen there are none. To flesh the point out, two bodies of research have provided relevant evidence. The first is exemplified by investigations such as Osborne (1980) and Watts (1983, reported by Gunstone and Watts, 1985). Following a methodology relatable to my 1991 study, these investigations involved pictures, often line drawings, around which pupils were interviewed. Typical of the pictures were a ball rolling on a table, a bicycle freewheeling upon a level road and a sledge skidding on some snow. Pupils were asked to indicate the forces operating at various points in the event and to justify their answers. Osborne's research was carried out with New Zealanders aged 11 to 14, Watts' with Londoners aged 13, 14 and 17. In both cases, there was a strong tendency to predict forces in the direction of motion at all points until the objects had come to rest. Because the pictures seldom depicted external sources of force, these forces were usually construed as internal 'impetus' acquired at initial propulsion. This again is reminiscent of part I.

The second body of research involved pupils moving objects themselves rather than merely reflecting on decontextualised events. Mention can be made here of the doctoral research of Lawson (1984, reported by McDermott, 1984), which had undergraduate students making a dry ice puck move at constant velocity across a smooth glass table. Movement was to be effected via a blast/blasts from an air hose. In this virtually frictionless environment, a single short blast would have done the trick but only about half the students realised this. The strategy of the others was to apply a continuing blast of constant force in the direction of motion, a continuing blast of diminishing force in that direction or even a series of short blasts. Encompassing an age range of greater relevance to this book, Langford and Zollman (1982) conducted a similar study with 11-, 14-, 16-, and 18-year-olds. Twenty-nine individuals were involved with roughly equal numbers at each age level. After practising with a set-up identical to Lawson's, participants worked with a computer simulation. This ensured a completely frictionless environment. When asked 'to make the puck move in a straight line across the table at a constant speed' (Langford and Zollman, 1982: 2), twenty-two participants responded with a continuing blast of constant force applied in the direction of motion. The 18-year-olds were somewhat more successful than the younger pupils, with 43 per cent applying a single short blast as opposed to 14 per cent of the 11-year-olds and 12 per cent of the 14-year-olds. Moreover, the 18-year-olds were more inclined to learn from their mistakes. Nevertheless, the emphasis at all age levels was upon continuing force.

It is interesting that Lawson's work and Langford and Zollman's emphasised constant velocity, whereas Osborne's and Watts' utilised objects that were probably interpreted as slowing down. (I say 'probably' because the data from my 1991 work that were presented in Table 5.3 indicate that such matters should not be taken for granted.) The implication is then that constant velocity requires constant external force applied in the direction of motion. In the absence of such force, moving objects will not instantly come to rest because they acquire impetus at initial propulsion which constitutes a further force in the direction of motion. However as impetus dies out, they will begin to slow down. This implies, of course, a simple model of stopping: objects come to rest in the absence of impetus and external push. Is this, then, the model that children typically subscribe to? The work of Gair and Stancliffe (1988) that was discussed in part II provides indirect evidence that it may be. It will be remembered that in this work, pupils aged 11 and 12 were shown mechanical toys in operation, for example a matchbox car being pushed along, and asked if and how the words 'force' and 'energy' could be deemed to apply. As for 'energy', three 'frameworks' were identified for 'force', the first two animistic but the third mechanical. However, although mechanical the third framework was defined simply in terms of action, that is 'a force is a push or pull which makes things happen' (Gair and Stancliffe, 1988: 175). This suggests that the only forces children contemplate are those directly or indirectly relatable to pushes and pulls, certainly air blasts on pucks or impetus after propulsion but probably not friction or drag. In which case, the model of stopping outlined above would seem to be entailed.

External forces opposing motion

Gair and Stancliffe's work is suggestive, but there is a problem. If objects come to rest in the absence of impetus and external push, why did so many pupils respond to my curling scene with the prediction of undiminished speed? External push was clearly absent with a freely rolling ball and impetus should have been seen as dissipating. Thus, the prediction on the outlined model of stopping would have been slowing down. Perhaps then, some children do have an inkling of external opposing forces like friction or drag, even if this is bound up with force in the direction of motion. The research which addresses the issue most explicitly is reported by Stead and Osborne (1980, 1981). This research involved in-depth interviews with forty-seven pupils aged

12 to 16, followed by a paper-and-pencil survey to over 800 pupils of equivalent age. After direct questioning about the meaning of 'friction', the pupils were shown line drawings, a sample of which are shown in Figure 5.1, and asked whether friction would be present in the depicted events. The results showed friction to be a meaningful word for most, although not all, of the pupils. However, it was a word whose meaning was both restricted and confused relative to received science wisdom. It was restricted in that many pupils saw it only as a relation between two solid objects, one of which is moving. Friction between an object and, say, water was frequently denied, as was friction between two stationary objects. Friction was a confused concept in that sizeable numbers of pupils equated it with gravity, energy and/or electricity.

Stead and Osborne help us to understand what children take to be the semantics of the word 'friction', but their work does not make it clear whether friction will be a significant mechanism in understanding motion. Certainly, the instances of horizontal motion described in this chapter have been the seemingly paradigm cases of one solid object moving against another. However we cannot assume that just because the presence of friction will typically be recognised with such instances, its causal significance

Figure 5.1 Examples of the line drawings used by Stead and Osborne (1980, 1981)

will also be appreciated. From this point of view a paper by Twigger *et al.* (1994) is perhaps more revealing. The central theme of this paper is an interview study with thirty-six pupils aged 10 to 15. As with the Stead and Osborne research, the pupils were asked what the word 'friction' meant to them. However, in addition to this, they were also asked to predict and explain what would happen in contexts of horizontal motion when friction was operating. The main contexts were a pebble being kicked across gravel, concrete and ice, and a model railway being pushed along a track. Encouragingly for science education (though discouragingly for the simple model outlined above), the majority of pupils were said to call upon friction when offering their explanations. The figures were 61 per cent for the pebble example and 72 per cent for the carriage. Interestingly, Twigger *et al.* report modest age trends, namely 'the introduction of friction in students' reasoning around age 11/12' (Twigger *et al.*, 1994: 227).

Twigger *et al.*'s study is probably the most comprehensive developmental investigation of friction available in the literature. Nevertheless, there is other research addressing the theme, and it gives mixed support to what Twigger *et al.* claim. Most consistent with Twigger *et al.* is the work of Inhelder and Piaget (1958). Discussing the study described earlier in the chapter, Inhelder and Piaget identify a close association between stage of reasoning and model of motion. In particular, they claim that first- and second-stage children have no conception of forces that oppose motion. Affirming in effect the simple model of stopping that was outlined earlier in the section, they propose that for such children 'the motion stops of itself by extinction of force imparted by the initial push, by fatigue, or by a tendency to rest' (Inhelder and Piaget, 1958: 125). It is not until the third stage that friction and 'air resistance' (that is, drag) are thought to be acknowledged. Since the third stage is regarded as achievable around the age of 12, this squares exactly with what Twigger *et al.* propose. Against this however are the results of Saxena (1988) in a study conducted with 118 respondents in the late school to university age range. One item in an eight-item question-naire asked respondents to select from four alternatives the resultant force on a scooter moving at a constant thirty kilometres per hour and to explain their answers. The modal response was to identify the resultant force as operating in the direction of motion. Indeed, only 20 per cent of this much older sample said that it would be operating in the opposite direction, admittedly often citing friction.

Intrigued by the differences between Saxena and the other researchers and intrigued also by Gilbert and Watts' (1983) claim that the literature on force and motion as a whole provides little evidence for age trends, I tried via the hopscotch and curling scenes of my 1991 study to shed some light. After answering the preliminary questions about 'What will happen to the ball's speed?', the pupils were asked why they thought this. Subsequently, after proposing variables to explain why given balls would or would not go fast, they were asked why the variables were important. Following the reasoning detailed in part II, I assumed that both lines of questioning provided contexts where causal mechanisms would be viewed as the only constructive response. The pupils' answers bore this out: either they said that they did not know or they invoked a mechanism. In response to one line of questioning at least, the vast majority of pupils did the latter.

The mechanisms were varied, but nevertheless categorisable, I felt, into four main types. The first type, which was rare, was non-physical in nature, for example the view of one 6-year-old that 'The ball will stop when it gets tired.' The second type was relatable to the 'impetus theory' outlined earlier, and included all responses that referred only to internal but consumable properties. Examples include 'It'll stop when the roll is gone', 'It's losing force as it rolls', 'It loses the energy which you give it' and 'The pressure she throws it with will slow down as it goes further.' The third type of mechanism called upon variable (and sometimes intermittent) external forces which oppose motion. These were sometimes seen to operate in conjunction with the balls' internal properties and sometimes in contradiction. Examples here include 'The bumps and cracks in the stones will slow it down', 'If it rolls very fast, it'll just go up and keep rolling as it hits the bumps' and 'Ice doesn't have any bumps so it won't slow down.' This mechanism was the most frequently used by my sample, accounting for 48 per cent of the responses to the hopscotch scene and 36 per cent of the responses to the curling. The final type of mechanism called upon constant external forces which oppose motion and included 'The pavement's surface will slow it and there won't be much force', 'The resistance of the ice will slow it' and from one 15-year-old who had clearly learned her physics 'If there's the same amount of force, the golf ball will go faster because it's small, so there's less air resistance and lots of momentum because of the mass.'

Although the third and fourth types of mechanism both refer to external opposing forces, they clearly differ in character. Moreover, not only do they differ but the differences are crucial to making sense of the unorthodox ideas about speed change that, as we saw earlier, many of the sample appeared to hold. It will be remembered that to quantify these ideas the pupils were assigned scores of zero if they had no comprehension of speed change, anticipating constant speed or even acceleration; scores of one if they had partial comprehension anticipating deceleration after an interval; and scores of two if their comprehension was adequate. Using these scores, I looked at ideas about speed change as a function of causal mechanism. For purposes of this analysis, I placed the pupils who said 'Don't know' to the mechanism questions with the pupils who called upon non-physical notions in a single 'No awareness' category. The other mechanisms were kept separate. It turned out that there were statistically significant differences in scores as a function of causal mechanism (F for hopscotch = 6.92, df = 3,122, p<0.001; F for curling = 18.59, df = 3,122, p<0.001). The mean scores which are presented in Table 5.4 were compared by follow-up Scheffé tests.

With the hopscotch scene, the pupils in the 'no awareness' and 'internal

Table 5.4   Relations between mechanisms and predictions of how speed changes: horizontal motion (Howe, 1991)

	No awareness		Internal properties of balls		Variable external forces		Constant external forces	
	Mean	SD	Mean	SD	Mean	SD	Mean	SD
Hopscotch	0.64	0.67	0.72	0.51	1.02	0.22	1.17	0.62
Curling	0.63	0.56	0.74	0.45	0.09	0.29	0.86	0.65

properties' categories obtained significantly lower scores than the pupils in the 'variable external' and the 'constant external'. No other comparisons yielded significant results. With the curling scene, the pupils in the 'variable external' category obtained significantly lower scores than the pupils in any other category. Again no other comparisons produced significant results. The reason for the patterns lies, I feel, in the presence of obvious bumps and cracks in the hopscotch photograph (in the form of concrete between the paving stones) and the absence of such bumps and cracks in the curling. To pupils who subscribe to variable external forces, the former should necessitate slowing down while the latter should definitely preclude it. To pupils who subscribe to more primitive mechanisms, slowing down may or may not be anticipated with either scene depending on whether the balls are expected to 'get tired' or 'lose impetus' before they cover the course. To pupils who subscribe to constant external forces, slowing down will be anticipated with both scenes but whether it will be predicted to be immediate or delayed will depend on the exact level of physics knowledge. Remembering what is entailed by scores of zero, one and two, the means and standard deviations in Table 5.4 are in all cases consistent with this gloss.

Given Table 5.4, we can conclude that the distinction between variable and constant external forces does not simply make sense conceptually but also helps us understand some otherwise puzzling results. Despite this, the distinction has not previously been made in the literature. Knowing this, I began to wonder about the analyses of friction offered by Twigger *et al.* (1994), Inhelder and Piaget (1958) and Saxena (1988). Could it be, I speculated, that the studies differed over whether variable external, constant external or both were included in the concept of friction? If so, could this explain the discrepancy over age trends mentioned earlier? To ascertain the extent of age trends when variable external and constant external were differentiated, I categorised the pupils by age and mechanism as shown in Table 5.5. Taking the categories in Table 5.5 one-by-one, I conducted chi-square analyses on the numbers of pupils at each age level using the category or failing to use it. There were no significant age differences over no awareness, internal properties or variable external forces. However, the age distribution for constant external forces was highly significant for both the hopscotch scene ($\chi^2 = 12.65$, df = 4, p = 0.01) and the curling ($\chi^2 = 23.34$, df = 4, p<0.001). As can be seen from Table 5.5, the difference arose because the 14- to 15-year-olds stood apart.

INTERNAL FORCES, EXTERNAL FORCES AND VARIABLE SELECTION

The causal mechanisms that children invoke for moving and slowing down in a horizontal direction appear, then, to show some changes with age. In this respect, they mirror what we observed earlier for variable selection, for there too age-related changes were documented and as with mechanisms they occurred relatively late. In particular, with the curling scene, there was improvement with age in the usage of relevant variables. This improvement was accounted for by the 14- and 15-year-olds making enhanced use of smoothness as a facilitator of going fast. In addition, with both the hopscotch scene and the curling, there was an increase with age around 12 and 13 in the usage of heaviness as a facilitator of going fast. There were not, however, any age-related changes in the usage either of irrelevant variables as a group or of the individual irrelevant variables that were sufficiently 'consensual' to be analysed separately. The parallels between mechanisms and variables are far from perfect, but they do perhaps give us some grounds for anticipating an association. The aim of the present section is to see whether the association can be supported with evidence.

Parallel but separate development

When we looked for an association between mechanisms and variables in part II, our point of departure was to correlate mechanism 'scores' with numbers of relevant and irrelevant variables. In the present context, this would not be appropriate. In contrast to the previous chapters, the mechanisms in Tables 5.4 and 5.5 do not lie on an unambiguous continuum in terms of adequacy. It is true that 'constant external forces' are nearer the mark than 'variable external forces' and that both show better understanding than 'no awareness' or 'internal properties'. However which is best, to show no awareness or to attribute internal properties when they do not exist? Any decision here would surely be arbitrary, meaning that we do not have a continuum. Without a continuum, we do not fulfil even the minimum requirement for correlational analysis. In addition though, to restrict the analysis to relevant and irrelevant variables as a group would be to sideline the interesting patterns observed earlier when the variables were considered separately. However, since each individual variable could be used (score = 1?) or not used (score = 0?), separate analysis could not validly proceed via correlation.

Recognising these points, I decided to conduct two sorts of analysis on

Table 5.5   Mechanisms of motion as a function of age: horizontal motion
(Howe, 1991)

	6–7- year-olds	8–9- year-olds	10–11- year-olds	12–13- year-olds	14–15- year-olds
Hopscotch					
No awareness	5	3	1	0	2
Internal properties of balls	12	6	5	9	4
Variable external forces	8	15	15	11	12
Constant external forces	1	1	3	4	9
Curling					
No awareness	8	9	5	4	3
Internal properties of balls	5	4	4	6	5
Variable external forces	10	10	12	9	4
Constant external forces	3	2	3	5	15

my 1991 data: (a) ANOVAs to compare the number of relevant and then irrelevant variables cited by the pupils who relied on each of the identified mechanisms; and (b) chi-square to compare the use vs. non-use of each consensual variable by the pupils who relied on each mechanism. With the hopscotch scene, there were near-significant associations between mechanisms and both relevant (F = 2.42, df = 3,122, p = 0.07) and irrelevant variables (F = 2.57, df = 3,122, p = 0.06). There was in addition a highly significant association between mechanisms and the belief that heaviness aids going fast (χ^2 = 14.42, df = 3, p<0.01), though mechanisms were unrelated to the other consensual variables. Detailed analyses revealed that heaviness was called upon significantly more often by the pupils who subscribed to constant external forces (χ^2 = 13.36, df = 1, p<0.001). The other mechanisms had no bearing on the use of heaviness. With the curling scene by contrast, there were no significant or near-significant associations to be found. This was as true for relevant and irrelevant variables as a group as it was for consensual variables taken separately.

It is hard to interpret the discrepancy between the two scenes. One thing is sure though: it probably does not mean separation between the everyday physics of hopscotch and the everyday physics of curling. There were significant correlations between the numbers of relevant variables used with each scene (r = +0.26, df = 124, p<0.01) and the numbers of irrelevant variables (r = +0.31, df = 124, p<0.001). There was also a strong association between references to heaviness across the two scenes (χ^2 = 21.99, df = 1, p<0.001).

In view of such data, it seems likely then that the hopscotch and curling scenes were integrated in the pupils' thinking, and indeed that the same knowledge structures were being applied to both. However, were the structures theoretically organised? The signs are not encouraging. In the first place, it would be necessary to place more emphasis on the hopscotch results than the curling to argue the case, and it is hard to see how this could be justified. In addition, the associations between mechanisms and variables were not exactly impressive for hopscotch, even if they were better than for curling. The only significant association was, after all, the one relating to heaviness. Nevertheless, with all this recognised, it still seemed worth scrutinising the hopscotch data to see first whether there was a dependent relation between mechanisms and variables and second whether any such relation was indicative of theorising.

It was noted of course that if dependency existed in the hopscotch data, it would not be in the sense observed in part II with the transfer of heat. There, every change in mechanism knowledge was related to (and arguably causative of) corresponding changes in variable selection. With the hopscotch scene, this could not possibly have been the case, for the pupils in the 'no awareness', 'internal properties' and 'variable external' categories were indistinguishable over the variables they chose. The only chance of a dependent relation lay with the pupils who subscribed to constant external forces, for they did differ from the other pupils over variable selection. Unfortunately, closer inspection of the data provided no grounds for thinking that dependency on mechanisms was occurring in practice. It was as possible for the pupils to pass thresholds of relevant and irrelevant usage without continuous external forces as it was for them to do the reverse. As for heaviness, thirty-five pupils cited it without subscribing to constant external forces while only four did the reverse. This suggests that if there was any dependency, it was mechanisms upon variables.

The picture is, then, as different as can be imagined from the transfer of heat. Associations between variable selection and causal mechanisms were only observed with one of the two scenes. However even with that scene there was nothing to suggest that variable selection depended on mechanisms. Indeed apart from the heaviness data, there is nothing with that scene to suggest any form of structural relation between variables and mechanisms. Perhaps then, this is what we have, two independent structures both of which improve in orthodoxy over time (and therefore show correlated development) but neither of which bears on the other. Although hard to reconcile with part II, and very troubling for our theory-based approach, two studies, both conducted by research groups centred on the Open University, suggest that this is precisely what happens with horizontal motion.

The two Open University studies involved computer simulations which allowed learners to explore motion in a horizontal direction and to manipulate its parameters. The first study, reported by Spensley *et al.* (1990), utilised an interface whereby 'blocks' of differing heaviness and surface area could be dropped from varying heights. After being dropped, the blocks would emerge from a 'chute' and from there travel in a horizontal straight line along 'surfaces' of varying slippiness (that is, ice, sand or syrup). The task was to ascertain the determinants of distance travelled. The demand characteristics of the interface were therefore ones in which variables were highlighted, respectively heaviness, surface area, height and slippiness. This is the major, and for us crucial, contrast with the second study which is presented in Hennessy *et al.* (1995a, b), for here the emphasis was squarely on mechanisms. In particular, the interface invited learners to explore the amount of force needed to propel groceries along floors, skaters along ice and speedboats through waves. Variables could be manipulated, for example the slippiness of the floor under the grocery box and the build of the skater. However, forces (and particularly forces in the direction of motion) were paramount, and were in fact the explicit focus of the worksheets that organised the learners' activities.

Spensley *et al.*'s report is limited to informal observations of small groups as they worked with the interface. The groups comprised Open University students who were attending a Summer School. Despite the informality of the data, it is clear that learner dialogue and problem-solving success were restricted to variables. Groups kept within the parameters of the problems 'and did not pursue the reasoning behind their initial expectations' (Spensley *et al.*, 1990: 11). Moreover, some proceeded by 'ignoring the issue of friction completely' (Spensley *et al.*, 1990: 14). Hennessy *et al.*'s approach to evaluation was considerably more structured, involving paper-and-pencil tests to the twenty-nine 12- to 13-year-olds who participated in the study. Tests were administered prior to small-group work with the interface as well as afterwards and involved the prediction of outcomes and the explaining of predictions. Test items were included relating to variables, for example understanding that heavier things are harder to get going and stop going, and to mechanisms, for example, friction is a force which acts in a direction opposing motion. Post-interface, participants had significantly superior ideas about force. They were, for example, less likely to exclude friction from their accounts or to claim that motion entails a continuing force. However, there was no change with any of the items relating to variables.

The speed and speed change systems

The point is that with both of the Open University studies, the component of knowledge that was highlighted progressed but the background component remained untouched. This implies, then, that knowledge structures relating to horizontal motion are highly compartmentalised, with variable knowledge encapsulated apart from mechanism. What does this mean? Without doubt, it means bad news for our theory-based approach, for it signifies variables beyond the scope of mechanisms when such variables would be denied. However, does it necessarily mean good news for the action-based approach in either its Piagetian or Vygotskyan guise? Certainly, the action-based approach in both guises could deal with outlying variables. Indeed, it goes so far as to predict these. Nevertheless, would it not also predict some tie-up between mechanisms and variables? After all, the encapsulation documented in my 1991 study continued up to the age of 15. Spensley *et al.*'s study related to adults. Yet even Piaget believed that by 11 or 12 children attain the 'possibilistic' perspective which allows them to ask 'Which variables are entailed by my mechanisms?' and 'Do my mechanisms entail the variables which I currently endorse?' Given that mechanisms were firmly in existence, some breakdown of the encapsulation would surely be predicted around 11, albeit a partial one for some years to come.

Breakdown would be predicted, but only if the variables and mechanisms related to the same kind of event. It is difficult to be certain about the Open University research, but in my 1991 study this was probably not the case. In the study, the pupils were asked about the variables relevant to *overall* speed, via questions about why selected balls would cover the course more quickly. They were given every opportunity to follow this up with mechanisms relevant to overall speed, via questions about why the nominated variables were significant. Although not recorded earlier in the chapter, these questions were overwhelmingly greeted with 'Don't know'. The mechanisms which the pupils generated in abundance were almost always given in response to the other trigger, namely the 'Why?' follow-up to 'What will happen to the ball's speed?' As we saw earlier, the mechanisms were in fact highly predictive of what the pupils envisaged happening to speed. However, the question 'What will happen to the ball's speed?' is of course a question about speed *change*, and not overall speed. This made me wonder whether the pupils had knowledge structures which made sharp distinctions between speed and speed change, the former constrained by atheoretically generated variables and the latter explained by causal mechanisms (which might or might not generate other variables – the 1991 study has nothing to say here). Certainly, the action-based approach could tolerate variables which were atheoretical in the absence of appropriate mechanisms. There would only be problems with variables which were atheoretical (in their entirety and after the age of 12) when appropriate mechanisms had been derived.

Could the encapsulation of variables and mechanisms really hang on the speed/speed change distinction? Certainly, the alignment cannot be dismissed as a methodological artefact. As noted in part II, the 1991 study asked about heat transfer mechanisms in two contrasting contexts which were exactly equivalent to the two used with propelled motion. With heat transfer, the contexts were indistinguishable in terms of the number and form of the mechanisms generated. In addition though, the material with which the present chapter began signals a developmentally plausible story of how speed and speed change might have created the encapsulated structures. The material in question is the research conducted or stimulated by Piaget

(1970) into how speed is computed. Its message, it will be remembered, is that instead of referring to $v = d/t$, children consider overtaking and finishing.

The key point is that if children consider overtaking and finishing, they cannot be looking at motion as it extends over time. Without information about extension over time, they are precluded from direct analyses of how speed changes. Precluded from direct analyses, they must unearth mechanisms to tell them the answer, hence the mechanism–speed change association. As we saw earlier, Newton linked speed change with speed via the formula $v = v_o + at$. However, there will be few embryonic Newtons in the school-aged population, so we can safely assume that the linkage escapes general detection. As a result, the speed change mechanisms will not be brought to bear on speed, meaning that we will either have speed, speed mechanisms and speed variables or speed, speed variables and optionally speed mechanisms depending on whether the theory- or action-based approach is closest to the mark. At which point, we can return to the section's data, and suggest that everything is currently pointing to an action-based gloss.

The chain of causality is both hypothetical and post hoc, and its only saving grace is that it concurs with the data. However, if it is correct, we should find encapsulation of variables and mechanisms in other contexts where speed is an issue. It is hard, after all, to imagine that children judge speed (and hence speed change) in one way in the contexts considered so far, but in a completely different way in contexts which contrast. Thus, it ought to be possible to firm the story up by considering other forms of motion. This is what the next chapter will attempt to do.

6 Horizontal and vertical motion compared

The previous chapter concentrated on the simple situation where objects propelled in a horizontal direction continue to move in that direction along a surface. The chapter unearthed a range of variables that children regard as relevant to the speed of motion. It also identified a series of mechanisms, of varying adequacy, which children refer to. The chapter showed that understanding of both variables and mechanisms improves with age. However the improvement is not indicative of a dependent relation. On the contrary, all the evidence pointed to two encapsulated systems, developing in parallel but completely autonomous. Seeking to explain this, the chapter made reference to the primitive notions of speed that children seem to hold, notions that rely on overtaking and stopping position rather than distance per unit time. The point was made that such notions preclude a direct analysis of speed change, rendering it instead a matter of inference. It was proposed that mechanisms are seen by children as the means to such inference, and certainly there was a close tie-up between beliefs about mechanisms and judgments of change. However, if speed and speed change are separated and mechanisms are seen to bear on the latter, speed must operate as an atheoretical construct. The same must therefore apply to its determining variables.

The previous chapter acknowledged that its use of the distinction between speed and speed change was hypothetical and entirely post hoc. However, it also acknowledged that it possessed an important advantage, the possibility of prediction. Since it is difficult to envisage children judging speed and speed change in one way in some contexts and in a different way in others, the implication is that variables and mechanisms will be encapsulated in all contexts where speed is an issue. Thus, by looking at other contexts, it should not simply be possible to firm up on the relevance of the speed/speed change distinction. It should also be feasible to obtain further evidence on the encapsulation of variables from mechanisms, a matter of the greatest importance given the concerns of this book. Thus, further research relating to speed is certainly warranted, but within which contexts? As noted, the previous chapter centred on horizontally propelled objects which sustain their initial direction, and the obvious contrast would appear to be horizontally propelled objects which do not do this. Instances of the latter include the bowled ball in cricket, the bullet in riflery, the toboggan in sledging, and the thrill-ride in Disneyland, instances where objects move downwards because they are not supported by a surface or because the surface is a slope.

On the face of it, downwards motion after horizontal propulsion would seem to provide the ideal point of contrast for the purpose we have in mind. Nevertheless, we need to be careful. It is clear that contexts involving downwards motion after propulsion differ over the attention that is paid to the downwards direction. It is paramount in Disneyland but less so in cricket, for when we designate someone a 'fast' bowler, we refer to the time that his/her ball takes to cross the pitch, and not to drop to the ground. When the downwards motion is discounted, it may mean that subjectively the situation is no different from what we have discussed already. Recognising this, it would be unsafe to proceed with downwards motion after propulsion without careful consideration of how such motion is conceptualised in the thinking of children. This is how the present chapter will start. Evidence will be presented which suggests that some instances are treated in a fashion which is analogous to simple horizontal motion. Others, however, are assimilated to vertical fall, indicating that the latter and not horizontal propulsion should become the focus. Accepting this, the chapter will continue with its usual analysis of variables, mechanisms, and variable–mechanism relations.

DOWNWARDS MOTION AND VERTICAL FALL

The literature is unclear about how adults conceptualise downwards motion after horizontal propulsion, let alone the situation with children. For instance, Hayes (1979) lays out a 'naive physics manifesto' in which he claims that 'vertical gravity is a constant fact of life, so vertical dimensions should be treated differently from horizontal dimensions', that 'there is a distinction which seems crucial between events which can just happen and events which require some effort', and that 'there may be two distinct ways of conceptualising motion: as a displacement or a trajectory' (Hayes, 1979: 256, 259 and 261 respectively). The first claim suggests that all events involving downwards motion will be differentiated psychologically from simple horizontal motion, the second that this will only apply with vertical *fall*, as opposed to, say, tossing, and the third that the crucial distinction is horizontal motion and rolling down slopes vs. propelled and vertical fall. Perhaps, though, all the potential categories are conceptually distinct, for Hayes also states 'If an object is moving, there are only five possibilities. It may be falling; or it may be being pulled or pushed by something; or it may be moving itself along; or it may be sliding; or it may be rolling' (Hayes, 1979: 261). Bliss *et al.* (1989) have applied Hayes' work in research with children, and they have also noted the above uncertainties. They express particular puzzlement about rolling down slopes, and suggest that further analysis of the whole field is in order. Such analysis will occupy us for the next few paragraphs, in order to see if and where there is meaningful contrast with the theme of the previous chapter. Having identified vertical fall as a promising focus, the discussion will then move to an analysis of variables.

Children's conceptions of downwards motion

We shall start with the situation which featured prominently in part I, falling after propulsion into space. When this happens, the objects' velocity will, as discussed in part I, decelerate in the horizontal direction and accelerate in the vertical direction. The deceleration is due to friction and drag, and the acceleration to gravity. It is this combination of horizontal deceleration and vertical acceleration that is responsible for the parabolic paths in the forwards direction which, as we saw in part I, objects will follow as they move through space. Thus, it is possibly ignorance of this combination that underlies the research results presented in part I. These results, it will be remembered, stemmed from the work of McCloskey (1983a, b) and the follow-ups it inspired. They amounted to widespread failure on the part of undergraduate students (indeed students of physics) to anticipate parabolic paths, the preferred responses being vertical straight lines, diagonal straight lines and right angular straight lines. Children experience similar problems, for two studies using in each case a single McCloskey-style problem have documented frequent anticipation of vertical straight lines.

The first study was mentioned in part I. It was reported by Eckstein and Shemesh (1989) and involved participants predicting and explaining whether a ball falling from a pole attached to a moving cart would land in a cup placed directly below. Although Eckstein and Shemesh's sample included adults, school-aged children were also involved. In fact, the school-aged group comprised 270 pupils aged 9 to 16, of whom 159 had not received relevant teaching. As it turned out, the responses of the 'untutored' and 'tutored' subsamples scarcely differed. In both cases, the majority could be placed into the two categories identified in part I: predicting that the ball would fall into the cup because of their quasi-magnetic relation or predicting that the ball would miss because it would fall straight down.

Closely related to Eckstein and Shemesh's research is a study by Marioni (1989) where participants were asked to imagine themselves running with a marble. The question was where would they have to release the marble to hit a target upon the floor. Marioni's sample consisted of sixteen 10- to 11-year-olds, twenty-five 12- to 13-year-olds and twenty-seven 16- to 17-year-olds. Only seventeen of these individuals gave the correct response of 'before the target', with the majority coming from the 12 to 13 age group. 'Above the target' indicative of an anticipated vertical fall, was the most frequently given response, accounting as it turned out for all but two of the youngest age group. Although 'above the target' was also popular with the 16- to 17-year-olds, their modal response was 'after the target'.

By demonstrating the popularity of the vertically downwards response, Marioni is endorsing Eckstein and Shemesh's main finding, that many children ignore the horizontal component when dealing with falling objects. However, he is also going beyond Eckstein and Shemesh in showing that there are at least two other paths that children acknowledge. I say 'at least two' because the pupils in Marioni's sample could have been contemplating any number of different routes from before/after the target to upon the floor. Whether any of these corresponded to a parabola is one moot point, and this seems to be one area where further information is required. However, it is not the only one, and one issue in particular is highly relevant to our present concerns. To elaborate, Marioni and Eckstein and Shemesh's focus on a single problem allows something (probably not everything) to be said about the variability across children. However, it does not permit inferences to be drawn about the variability across problems, and yet this is the crucial issue when trying to establish conceptual categories.

McCloskey is as guilty as Eckstein and Shemesh and Marioni of not paying attention to across-problem variability. However, it is clear from McCloskey's results that his problems differed in the proportion of vertical, diagonal and right angular responses that they evoked. Since the problems were solved by the same individuals, this signifies across-problem variability within individuals even if the variability was not systematically explored. As such, McCloskey's work was a direct trigger for a little-known study by Maloney (1988) which looked in some depth at how individuals vary across a problem range. Maloney's study involved sixty-four undergraduate students who were neither taking, nor had taken, a college-level physics course. The students were presented with a series of problems, half of which involved a ball being thrown off a cliff and half a stream plummeting into a waterfall. Their task was to predict either the distance travelled or the time to fall under different combinations of object weight, pre-fall velocity and cliff height. Maloney's quantitative analyses attest to the variation of prediction with all three variables. However, what seems particularly significant is the comment made by one student to the effect that 'If two balls of different weight are thrown off equal height cliffs, then the lighter ball will go farther. It seems to me that a heavy ball will head for the ground a lot faster than a lighter ball – due to gravity, and the longer the ball is in the air "arching" the farther it will go' (Maloney, 1988: 507). What struck me about this line of reasoning was not simply the clear reliance on weight but also the claim by undergraduates that heavy balls resist motion in the horizontal direction, for this is exactly the reverse of what the Howe (1991) data have suggested so far. It will be remembered that with the hopscotch and curling scenes discussed in the previous chapter, there was a strong tendency for pupils aged 12 and over to argue that heavy objects travel faster than light ones in a horizontal direction. If anything, it was the pupils under 12 who saw heavy objects as resistant to horizontal motion.

Unwilling to hypothesise a curvilinear relation whereby children under 12 and adults are differentiated from teenagers, my first reaction was to dismiss Maloney's student as an isolated case. However, I was well aware of a body of evidence demonstrating that when university students are questioned about vertical fall without horizontal motion, they typically say that heavy objects fall faster than light ones. The evidence stems from research by Champagne *et al.* (1980a), Gunstone and White (1980, 1981) and Whitaker (1983) where simple apparatus involving two suspended objects was either presented to students or described to them. The objects were clearly differentiated in terms of weight, and the students' task was to predict which would reach the ground first if both were allowed to fall. Frequently, no difference between the objects was predicted. However when a difference was predicted, it was almost always in terms of the heavy object being first.

Knowing the work of Champagne *et al.*, Gunstone and White, and Whitaker, I began to wonder whether Maloney's data might not be indicative of problems involving fall under horizontal motion being assimilated to problems involving simple vertical fall. This would mean that of all Hayes' (1979) claims the one about displacements vs. trajectories would be closest to the mark, making in effect for an everyday physics of supported motion vs. an everyday physics of object fall. Given such a separation, the variables facilitating horizontal speed in contexts of object fall would not be the ones discussed earlier in this chapter but rather the ones inhibiting vertical fall. This would of course make object fall the ideal candidate for contrastive analysis with the previous chapter.

The possibility seemed sufficiently intriguing and important for my colleagues and I to attempt a partial replication and clarification of Maloney's research with 12- to 15-year-old pupils. By working with this age range, we would have a sample exactly comparable to the pupils in my 1991 study who had favoured heaviness as a facilitator of motion along a surface. In partial replication of Maloney's research, we varied horizontal velocity and object weight in problems which in all cases related to projected motion through space. Thus, we designed problems where the horizontal velocity could reasonably be described as fast, for example a bullet and an express train, and problems where it could reasonably be described as slow, for example a drifting ship and a putted golf ball. We produced eight fast horizontal velocity problems and eight slow ones. Within each set of eight, we had four problems where the falling object could reasonably be identified as heavy, for example a cable car and a treasure chest, and four where it could reasonably be identified as light, for example the bullet and a beer can.

As noted above, the study was a partial, not a complete, replication of Maloney's research. Where it went beyond Maloney was in embedding every combination of fast and slow velocity and light and heavy object in each of four contexts of pre-fall motion. Two of these contexts involved linear motion; and two involved non-linear. Examples of each context are shown in Figures 6.1 and 6.2. In the first linear context, the motion was also 'transparent', that is the object was clearly moving with respect to the obvious frame of reference. Examples include a golf ball rolling over a cliff and a ski jumper whizzing off a precipice. In the second linear context, the pre-fall motion was disguised, that is the objects were being carried by something that was moving. Hence the objects were not themselves moving with respect to the obvious frame of reference. Examples here include famine relief being dropped from an aircraft and pirate treasure being dropped from a galleon. In the first non-linear context, the pre-fall motion was pendular, for example a conker falling from its swinging string and a fairground artiste falling from her flying trapeze (safety nets being mentioned). In the second non-linear context, the pre-fall motion was circular, for example sparks flying off a Catherine Wheel and water coming out of a coiled hose.

We administered the problems in a paper-and-pencil test to 180 Glasgow pupils aged 12 to 15. Details of the test have been published in a number of sources, for example Anderson *et al.* (1992), Howe *et al.* (1991) and Tolmie and Howe (1993). In brief, it involved presenting the problems on worksheets, via a combination of text and pictures, and asking the pupils to complete the pictures by drawing the paths that the objects would take from starting to fall until hitting the ground. The responses that the pupils gave were scored along two dimensions. The first related to the predicted direction of motion, that is forwards, backwards or something else. The second related to the predicted shape of path. This dimension encompassed much that was included in the first but went beyond it. Moreover, scores along it were correlated +0.56 (df = 178, p<0.001) with scores along the

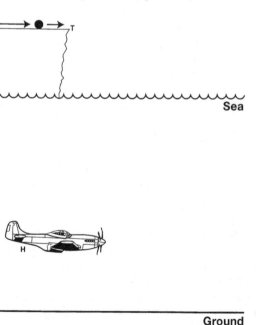

Figure 6.1   Examples of linear motion used by Anderson *et al.* (1992)

Safety Net

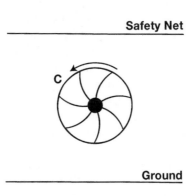

Ground

Figure 6.2 Examples of non-linear motion used by Anderson et al. (1992)

first dimension. Thus, for present purposes, the shape of path dimension is all that we need be concerned with. Accepting this, shape of path scores could vary from zero to five. A score of zero was given if the path was drawn vertically downwards; recognition of the horizontal dimension guaranteed something better. A score of one was given if the paths diverged from verticality in an esoteric fashion, for example steps made up of wispy curves. Paths that went outwards before going downwards were given scores of two, and paths that went diagonally downwards scores of three. Approximations to parabola were scored four or five depending on their adequacy.

Analysis of the scores revealed differences between pupils in how they responded to given problems and differences within pupils in how they responded across the problem range. The differences between pupils can be dealt with summarily because they are not central here. Essentially, there was a significant improvement with age (F = 5.55, df = 2,176, p<0.01), with the scores varying from the mean of 2.46 obtained by the pupils in the 12 to 13 age group to the mean of 2.80 obtained by the pupils in the 14 to 15 age group. More important is the fact that the problems differed in the responses they evoked. The differences were not, as it happened, a function of pre-fall motion. The mean scores for transparent linear, disguised linear, pendular and circular were all very similar. Rather, they lay in a significant interaction between horizontal velocity and object heaviness (F = 24.76, df = 3,522,[1]p<0.001). Regardless of pre-fall motion, the pupils were likely to give more accurate responses with fast, heavy objects (mean = 2.69) and slow, light ones (mean = 2.92) than with fast, light objects (mean = 2.53) and slow, heavy ones (mean = 2.41). The reason for the difference quickly became clear. Fast, light objects were particularly likely to elicit the outwards before downwards path which, as explained above, obtained a score of two. Slow, heavy objects were particularly likely to elicit the vertically downwards path which was associated with a score of zero.

Remembering the words of Maloney's (1988) student, it seems then that with this sample of 12- to 15-year-olds heavy objects will indeed 'head for the ground'. Moreover, in doing so, heavy objects will inhibit motion in the horizontal direction. Earlier it was proposed that when adults hold this belief, it may be because they assimilate events involving falling after horizontal motion to a general notion of falling which contrasts with an alternative notion of supported motion. Is the same true of children? Certainly there is evidence suggesting that it may be the case with the 12 to 15 age range. Firstly, research by Champagne *et al.* (1980b) indicates that the beliefs of 13- to 15-year-olds about falling in the absence of horizontal motion correspond to what we have already seen with adults: if weight is believed to be relevant, it is almost always the relatively heavy objects that are believed to fall fastest. Thus, there is parallelism for children just as for adults between propelled and non-propelled fall. Secondly data collected for a study reported by Howe *et al.* (1992b) show that when there is supported motion downwards because objects are rolling down slopes, heavy objects are not seen by 12-year-olds as inhibiting horizontal motion. On the contrary (and equivalently of course to horizontal motion without descent), they are seen as facilitative.

In brief, the Howe *et al.* (1992b) study involved the questioning of 113 pupils aged 8 to 12 about the variables relevant to motion down an inclined plane. There were roughly equal numbers of pupils in the four age cohorts involved, 8-to 9-year-olds, 9- to 10-year-olds, 10- to 11-year-olds and 11- to 12-year-olds. Some 102 pupils not only thought that object heaviness was relevant to motion down the plane but also were highly consistent in how they thought this would operate. In other words, they said either that 'Heavy objects will travel faster' or that 'Light objects will travel faster' on at least seven of the eight occasions they were questioned. It turned out that 33 per cent of the 8-to 9-year-olds in the subsample of 102 chose heaviness as the facilitating factor over lightness. A total of 28 per cent of the 9- to 10-year-olds did this, as did 24 per cent of the 10- to 11-year-olds. However, the figure for 11 to 12-year-olds was 67 per cent. This made for a highly significant age effect (χ^2 = 11.66, df = 3, p<0.01).

It seems then that by the age of 12 children have a clear 'falling objects' structure, which encompasses objects that are both moving and stationary in the horizontal direction. Combining the Howe *et al.* (1992b) data just presented with the conclusions of the previous chapter, it seems that the 'falling objects' structure contrasts with the 'supported motion'. Interestingly, this would square neatly with Bliss *et al.*'s (1989, see also Bliss and Ogborn, 1993) proposal that the golden rule of everyday physics is that 'If an object is not supported, it falls until it is once more supported' (Bliss *et al.*, 1989: 262). The notions endorsed by children under 12 are less clear. Below 12, lightness was facilitative of motion down an incline for the Howe *et al.* sample, and the previous chapter provided some, albeit limited, evidence for similar beliefs at this age level regarding motion along a horizontal surface. However, I do not know of any incisive research which considers the beliefs of the under-twelves regarding object fall. There is not, to my knowledge, anything on fall with horizontal motion or fall without. Obviously this gap is unfortunate but nevertheless the results with the over-twelves do point to what was anticipated earlier from the work with adults. To obtain a genuine contrast with the situations considered in the previous chapter, the context to work in is object fall regardless of motion in the horizontal direction. The question to ask is what inhibits or facilitates the downwards speed. Object fall is, then, what we shall be focusing on in the paragraphs to follow.

Variables relevant to object fall

By virtue of the preceding discussion, we have already established something of relevance to the issue of facilitatory and inhibitory variables as regards downwards speed. After all, the research that my colleagues and I conducted relating to parabolic paths revealed that heaviness is seen by 12- to 15-year-olds as facilitating downwards fall. Lightness is seen as inhibitory. The work of Champagne *et al.* (1980b) shows the same with vertical fall. However, is this all there is to say? So often in the preceding pages, we have found that children who rely on a handful of variables in controlled investigations that manipulate those variables only, open up and reveal a broad spectrum of ideas when the methodology is more informal. Thus, an informal investigation of object fall is certainly warranted. It is indeed also much needed because there is nothing of relevance amongst published research. As a result, we are once more thrown back on my 1991 study, for this study did include two items where the pupils were quizzed informally about the variables relevant to vertical fall.

In designing the items, I took account of some further implications of the drag force which as we saw in the previous chapter operates with motion through fluids. Drag, it will be remembered, is definable as $CpAv^2/2$, and is thus directly proportional to object velocity. When objects fall, their velocity increases due to the accelerating force of gravity. Hence, the drag force derived from air resistance also increases. If objects could fall indefinitely, a point would be reached where the drag force was equal to the downwards force from the object's weight. With the forces balanced and hence no net force, moving objects will according to Newton's First Law move with constant velocity. Thus, thanks to drag, there is in theory some point at which every falling object will attain a 'terminal velocity' and acceleration will cease. This is the theory but of course in the real world objects cannot fall indefinitely. Sooner or later they will hit the earth's surface. Thus, in producing items for my 1991 study, I selected one context where terminal velocity is typically achieved before hitting the ground and one context where terminal velocity is typically not so achieved. The contexts related respectively to parachutes and to the five balls that featured with the hopscotch and curling scenes as described in the previous chapter. These balls, it will be remembered, were a large green ball, a tennis ball, a table tennis ball, a golf ball and a bowls wood.

For the parachutes context, I prepared a photograph depicting a toy Action Man being lowered via a plastic parachute with three alternative parachutes in the vicinity. As before, the scene was associated with a set of questions aimed at eliciting beliefs about variables. The questions were as follows: 'The best parachute would be one that fell very slowly. Could anything else be done to the parachute that the toy is holding to make it fall more slowly? What? Why? Look at the other parachutes on the floor. Do you think that any of them could be used to make a better parachute? Which? Why? What about the others? Would they all be equally bad or would some be better than others? Why?' For the five balls, I utilised a photograph of the Leaning Tower of Pisa. During the interviews, it was explained that during the Middle Ages Galileo had thrown balls off the tower to resolve a controversy about speed of falling. The pupils were invited to imagine Galileo throwing the five balls that had recently been encountered with the hopscotch and curling scenes. They were then asked: 'Would some of them have fallen faster through the sky than others, or would they all have fallen with the same speed? Why? Which ball would have fallen the fastest? Why? What about the other balls? Would they all fall equally slowly, or would some fall faster than others? Why?'

As might be expected, heaviness featured prominently in the pupils' responses. Fifty-three of the 126 pupils said that a light parachute would fall more slowly and eighty-seven said that a heavy ball would fall more quickly. Only seven pupils thought that a slow speed would be associated with a heavy parachute and only twelve thought that a fast speed would be associated with a light ball. This is rather different from what was observed with the hopscotch and curling scenes, for there both lightness and heaviness were commonly associated with rapid speed. It is also interesting in that while lightness and heaviness were both partially relevant with hopscotch and curling, only heaviness could be said to be relevant with parachutes and the leaning tower. This is because when objects are dropped there is no force and hence no acceleration prior to propulsion. After dropping, there is as always friction and drag, and, as we know, the latter is indirectly related to heaviness. Hence, the heavier the object, the less the drag and the less the deceleration from that source. Thus, to mention heaviness as a facilitator of vertical fall is to say something relevant, and in this context it is noteworthy that such mentions increased with age with the leaning tower scene (χ^2 = 12.35, df = 4, p<0.05). While 54 per cent of the 6- to 7-year-olds called upon heaviness and 52 per cent of the 8- to 9-year-olds, the figures for the 10- to 11-year-olds, the 12- to 13-year-olds and the 14- to 15-year-olds were 83 per cent, 88 per cent and 70 per cent respectively.

There is no doubt that heaviness predominated with both the parachutes scene and the leaning tower, for no other variable was cited by more than 15 per cent of the sample. Nevertheless, other variables were cited, to the tune of thirty-seven different ones with the parachutes scene and ten with the leaning tower. With the benefit of hindsight, I feel that the leaning tower scene was not as successful as it might have been at eliciting the full variable range. A surprising number of pupils knew that the 'controversy' Galileo was addressing related to weight, and hence may have focused unduly on that dimension. This said, the leaning tower scene did generate nine other variables and with the parachutes scene there was an impressively broad range. With both scenes, some of the non-heaviness variables were scientifically relevant, for example smallness or smoothness as conducive to speed and bigness or roughness as working against. Others were considerably less satisfactory, for example the claims that a slow parachute must be plastic, real, soft, nice, cellotaped, tall, or leather and that a fast ball must be plastic, lucky, good, or round.

As before, I looked at usage of relevant and irrelevant variables as a function of age. With the leaning tower scene, there was a statistically significant increase with age in usage of relevancies (F = 2.45, df = 4,121, p<0.05) and a statistically significant decrease in usage of irrelevancies (F = 3.00, df = 4,121, p<0.05). Not surprisingly usage of relevancies was negatively correlated with usage of irrelevancies (r = -0.44, df = 124, p<0.001). The age effect reflected a steady increment or decrement, and thus despite the significant F values no single age group differed significantly from another. With the parachutes scene, there was also a significant negative correlation between usage of relevant variables and usage of irrelevant (r = -0.34, df = 124, p<0.001). However, despite a steady increase with age in the usage of relevant variables, neither relevancies nor irrelevancies changed significantly with age.

FORCE AND VERTICAL MOTION

Because the usage of relevant variables in my 1991 study was negatively correlated with the usage of irrelevant, we can infer that there was some incompatibility between the two. Thus, whatever precipitated the usage of relevant variables must have precluded the usage of irrelevant, or vice versa. This is interesting because although it does not necessitate causal mechanisms, it is certainly what would be expected if such mechanisms were making the running. Is this then what was happening in practice? We already know what kind of causal mechanism would render relevant variables necessary. It is one which postulates a downwards force akin to gravity opposed by an upwards force akin to drag and friction. Do some children subscribe to such a mechanism, and are these the children who focused on relevancies? The present section will attempt to find out.

Gravity and air resistance

As regards the first question, we have already discussed some pertinent material, and it is not encouraging. Take drag for instance. It occurs by virtue of passage through air. However, as we saw in the previous chapter, few children anticipate 'air resistance' or 'air pressure' when movement is horizontal. With friction, the work of Stead and Osborne (1980, 1981) offers an equally dismal message. There, it will be remembered, many 12- to 16-year-olds were found to restrict the notion of friction to solid objects. In fact, around 50 per cent of the sample was adamant that friction could not occur between an object and air. Yet against this is the research of Twigger *et al.* (1994) which, as mentioned in the previous chapter, was an interview study with thirty-six 10- to 15-year-olds. In addition to the items outlined earlier, there was a series of questions relating to parachutes. Line drawings were presented which depicted four phases in descent by parachute (from leaving an aircraft until landing safely). The pupils were asked to draw force arrows to explain the motion, and were then quizzed about the forces' magnitudes. Twigger *et al.*'s conclusions are straightforward: 'Air resistance as an upward force was used by all students' (Twigger *et al.*, 1994: 224).

Twigger *et al.* are equally bullish about gravity, and here other research seems to support them. For instance, as part of a study summarised in Osborne and Freyberg (1985), a line drawing was presented which showed the descent of a ball that had been tossed in the air. Some 500 pupils aged 13 to 17 were asked to indicate whether there would be an upwards force, a downwards force or no force. Although the results are not presented in detail, it is clear that at least 75 per cent of the sample must have opted for a downwards force. Also focusing on a line drawing of a descending ball was one item in the questionnaire used by Saxena (1988) in the investigation mentioned earlier. This investigation, it will be recalled, involved 118 respondents whose ages spanned the late school to university range. With the item of relevance, respondents were asked whether there would be a gravitational force, a push by hand, a magnetic force, a repulsive force and/or anything else (to be specified). A total of 77 per cent of the sample identified a gravitational force, with the remainder showing no particular preference.

On the face of it, the studies by Twigger *et al.*, Osborne and Freyberg and Saxena all indicate widespread subscription to gravity. Nevertheless, when we probe more deeply there are a number of problems. First, the terms 'gravity' or 'gravitational force' were only used explicitly in Saxena's work, and Saxena's age range is very much at the limits of our present concerns. As a result, there are difficulties in reconciling the three studies with research which paints a less rosy picture as regards references to gravity. An example of the latter is a study reported in the paper by Bliss *et al.* (1989) where twenty-six pupils whose ages must have ranged from about 11 to about 18[2] interpreted the falling depicted in comic cartoons. According to Bliss *et al.*, 'the idea (of gravity) is not very frequent' (Bliss *et al.*, 1989: 271). It accounted in fact for forty-four of the 292 interpretations offered. Second, there is no evidence that Twigger *et al.*, Osborne and Freyberg or Saxena explored the *nature* of the downwards force, irrespective of whether the word 'gravity' was used or not. This is unfortunate, for the work of Minstrell (1982) which was discussed in part I suggests that even students who have received formal instruction and therefore know the word 'gravity' perfectly well, are uncertain about how the force operates. They see it as varying with height, disappearing at floor level, and interacting strangely with weight.

As it happens, there is parallel evidence with untutored individuals, showing that even when they use 'gravity' to indicate a downwards force they seldom do so in a conventional fashion. Consider for example the studies by Noce *et al.* (1988) and Ruggiero *et al.* (1985). In the course of these studies, over 400 individuals whose ages ranged from early teenage to adulthood were questioned about what happens: (a) if an astronaut lets go of a spanner while standing on the moon; and (b) if the air is evacuated from a glass container in which a stone is resting. Many individuals believed that the weight of the spanner and stone would reduce in the absence of air, and the reasons they gave are highly revealing. In some cases, they saw air as a downwards force which creates weight in collaboration with gravity. In others, they saw air as a downwards force which creates weight through the mediation of gravity. In yet others, they regarded air as the meaning of gravity, arguing for example that gravity 'is a kind of air that pushes people down' (Noce *et al.*, 1988: 65).

The point is that for the samples studied by Noce *et al.* and Ruggiero *et al.* the concept of gravity was intimately (and erroneously) related to the concept of airiness. The samples were Italian, but cross-cultural research by Ameh (1987) and Watts (1982) has produced very similar findings. Ameh questioned four Nigerian teachers and their pupils about gravity in the contexts of standing on the moon, standing on the earth, and falling from a planet. Pupils and, alarmingly, teachers frequently made such claims as '(Gravity pulls) from the air around us' (Ameh, 1987: 213) and '(Gravity is) less than on the earth because there is no air there' (Ameh, 1987: 215). It was likewise for Watts in a study conducted in London. The pupils interviewed in the study spanned the British secondary school age range, that is 11 to 18. Across the age range, it was commonplace to hear that gravity is a force that requires air to act through and/or that where there is no air, there is no gravity.

Thus gravity and air can be interwoven in pupils' thinking, and can even be confused with each other. However, that is not all. As mentioned earlier, the work of Stead and Osborne (1980, 1981) suggests that gravity can also be confused with friction. Slightly over 20 per cent of Stead and Osborne's sample treated the two concepts as one. Knowing this, we must wonder of course about how air, gravity and friction are combined in children's thinking. Do some children integrate all three concepts into a general downwards force? Alternatively, do some treat two as a pair and the third as oppositional? Moreover, despite the work of Twigger *et al.* (1994), Osborne and Freyberg (1985) and Saxena (1988) do any children genuinely differentiate the three components into separate forces prior to formal teaching?

The study that I was involved in relating to parabolic paths (for example, Anderson *et al.*, 1992) gives some guidance as to how the above questions are likely to be answered. It will be remembered that the study centred around sixteen problems where 12- to 15-year-olds were asked to draw the paths that objects would follow as they fell through space. With four problems, there was a subsequent request to indicate (in words and by drawing) the forces that would be operating at the point of fall. As detailed in the various publications, proposed forces were scored on a scale from zero to six and great care was taken to avoid over-interpretation of what the pupils meant. Without wanting to go into details about what the scale points referred to, I can perhaps point out that it was not necessary to differentiate more than one relevant force to obtain a score of two. Two relevant forces and a score of at least four was guaranteed. Thus, it is instructive that across the sample as a whole, the mean score for the four problems was 1.13.

However, while it can be inferred from our data that relevant forces will go unrecognised, little can be said about which forces are likely to be included on the few occasions that this takes place. For that purpose we need more qualitative analysis and once again this is where my 1991 study can make a contribution. With the parachutes scene, the interviewing began with the question 'What will happen to the parachute's speed as it falls through the sky?' No matter what the pupils said, they were then asked why this would happen, a clear invitation to specify a mechanism. In addition, every variable generated in answer to the questions about better and worse parachutes was followed with a request to explain why the variable was important. As argued already, this can be assumed to carry the force of a request for a mechanism. With the leaning tower scene, there were no questions about speed change. This was a consequence of the schools' understandable preference for a maximum half hour's interview with each pupil, a constraint which resulted in the omission of several otherwise desirable features. However, the generation of variables with the leaning tower scene was always followed by the question 'Why is (variable) important?'

Twenty-seven pupils did not generate a mechanism for the parachutes scene and, probably reflecting the more limited opportunities for mechanism discussion, forty-four did not do this for the leaning tower. Amongst the remainder there was a clear division between mechanisms that were entirely erroneous and mechanisms that were partially correct. The former were of two types: inherent properties of the falling object or downwards push from the air. Thus, on the one hand, we had 'The hole in the top will keep it up' and 'The bowling ball will go fastest because there's more energy in it.' On the other, we had 'All the wind will go into the parachute and push it down' and 'The air pulls faster on heavy masses'. In fact, inherent properties were considerably more popular than downwards pushes. Twelve pupils generated the former for the parachutes scene and five the latter. The corresponding figures for the leaning tower scene were forty-one and three. These figures are of interest given the Twigger *et al.* (1994) study that was discussed earlier. Twigger *et al.* did not find any students referring to the downwards force of air during descent by parachute. However, what is non-existent in a sample of thirty-six could easily be existent but rare in a sample of 126.

As regards partially correct mechanisms, eighty-one pupils mentioned the upwards pressure of air with the parachutes scene and twenty-two did this with the leaning tower. Thus, we had 'The parachute's speed would deteriorate because there's more air pushing it up', 'The light ball would get caught up by air pressure and couldn't get through' and, impressively, 'In theory all the balls would come at the same speed – in practice it depends on the air resistance – Galileo would have been better to have done his experiment on the moon.' By contrast, one pupil mentioned the downwards pull of gravity with the parachutes scene and fifteen did this with the leaning tower, venturing remarks like 'The heavier ones will go faster because there's more gravity to pull them down' and 'Gravity tugs hard on big masses.' One child mentioned both air pressure and gravity with the parachutes scene and two did this with the leaning tower.

Because some mechanisms were used infrequently, there had to be a degree of cross-mechanism combination to permit valid analysis of age effects. The most sensible approach here seemed to be combination into no mechanism, erroneous mechanism and partially correct mechanism. The distribution that this achieved is shown in Table 6.1.

Chi-square analysis revealed a nearly significant association between mechanisms and age with the parachutes scene ($\chi^2 = 14.76$, df = 8, p = 0.06) and a highly significant association between mechanisms and age with the leaning tower ($\chi^2 = 24.60$, df = 8, p<0.01). Table 6.1 shows that with the leaning tower scene the 14- to 15-year-olds stood apart. Perhaps surprisingly, there was also a strong association between mechanism category with the parachutes scene and mechanism category with the leaning tower ($\chi^2 = 24.08$, df = 8, p<0.001).

Air, gravity and variable selection

So were mechanisms implicated in variable selection? As in the previous chapter, two analyses were conducted on the 1991 data to explore this

Table 6.1   Mechanisms of motion as a function of age: vertical motion
(Howe, 1991)

	6–7- year-olds	8–9- year-olds	10–11- year-olds	12–13- year-olds	14–15- year-olds
Parachutes					
No mechanism	8	7	5	4	3
Erroneous mechanism	7	5	1	2	2
Partially correct mechanism	11	13	18	18	22
Leaning Tower					
No mechanism	12	9	6	10	7
Erroneous mechanism	13	9	10	8	3
Partially correct mechanism	1	7	8	6	17

possibility: (a) ANOVAs to look at the usage of relevant and irrelevant variables as a function of mechanism category; and (b) chi squares to look at the usage of heaviness to predict rapid speed, again as a function of mechanism category. The two scenes produced markedly different results. With the leaning tower scene, there were statistically significant associations between mechanism category and both relevancies (F = 10.58, df = 2,123, p<0.001) and irrelevancies (F = 4.04, df = 2,123, p<0.001). Usage of heaviness (obviously the main source of relevancies) was also associated significantly with mechanism category (χ^2 = 17.90, df = 2, p<0.001). With the parachutes scene, none of the analyses produced significant results.

However while the results from the two scenes were different, the message was similar, for in neither case was there evidence that particular mechanisms were associated with particular variable types. On the contrary, follow-up analysis of the leaning tower data revealed that the pupils who subscribed to erroneous mechanisms and the pupils who subscribed to partially correct mechanisms were more or less equally likely to mention relevant variables. The reason for the significant results was that in both cases they were markedly more likely than the pupils who subscribed to no mechanisms to mention relevant variables. This was reflected in the references to heaviness which were made by 79 per cent of the erroneous mechanism pupils, 85 per cent of the partially correct but only 45 per cent of the no mechanism. This absence of a tie-up between particular mechanisms and particular variables was also true for irrelevancies. Here again there were no significant differences between the erroneous and partially correct pupils, although this time only the latter group differed significantly from the no mechanism pupils. As quantitative evidence for these points, Table 6.2 presents the mean numbers of relevant and irrelevant variables as a function of mechanism.

A consistent picture of encapsulated knowledge

The results just obtained are of course very similar to what was observed in the previous chapter with hopscotch and curling. With those scenes too, there was evidence, albeit limited, for age-related improvements in variables and mechanisms. Interestingly, the evidence also amounted to significant age effects with one scene but not with the other. However, despite the tie-up over age effects, there was nothing at any age level to indicate a contingent relation. There were in particular no signs of advances in variables that waited on advances in mechanisms, or indeed of the reverse. Noting these points, we are entitled perhaps to conclude that the results with the parachutes and leaning tower scenes parallel the results with the hopscotch and curling in all key respects. From some perspectives this must seem surprising. After all, as emphasised earlier, there appears to be a big distinction in everyday physics between supported motion and falling. The distinction is as true for children as it is for adults, and it is so powerful that when objects are projected into space, analyses of the horizontal component are entirely subservient to analyses of the vertical. This is why the present chapter which began with propulsion under horizontal velocity has devoted so much attention to simple fall.

The phenomenological differences between the scenes have had some consequences. The variables proposed for the parachutes and leaning tower scenes have differed from the variables proposed for the hopscotch and curling. The same has proved true for the mechanisms. However, the suggestion is that in the area of focal concern, the age profile over variables, mechanisms and the variable–mechanism relation, the picture is consistent. Why should this be?

Table 6.2   Relations between mechanisms and variables: vertical motion
(Howe, 1991)

	Relevant variables	*Irrelevant variables*
Parachutes		
No mechanism	0.41	1.26
Erroneous mechanism	0.59	0.94
Partially correct mechanism	0.63	1.09
Leaning Tower		
No mechanism	0.57_a	0.32_a
Erroneous mechanism	0.84_b	0.23_{ab}
Partially correct mechanism	1.03_b	0.05_b

N.B. Where subscripts differ within a column, the values were significantly different on a Scheffé test ($p < 0.05$)

In the previous chapter, the results with the hopscotch and curling scenes were interpreted as indicating two systems of encapsulated knowledge. The first system was deemed to be directed at overall speed, and by virtue of this to encompass the variables seen as relevant to travelling fast or travelling slowly. The second system was taken to focus on speed change, and was presented as the raison d'être for mechanisms. In particular, the 'perceived' mechanisms were regarded as the bases for judging acceleration, deceleration or constant speed. The role of mechanisms was seen to be necessitated by the criteria which children use for assessing speed, criteria which preclude direct perceptions of speed change. The two systems were proposed earlier with the utmost tentativeness, but they do gain support from the present section's results. Speed is a directionless, non-vectorial concept, and not just from the scientific perspective. The previous chapter began with the research of Jones (1983) which shows that this is how speed is construed by children (and also how, erroneously, velocity is construed). However, if the concept of speed is directionless, its consequences for horizontal motion should be mirrored with vertical. As such, parallels between the four motion scenes in the 1991 study are strongly predicted.

Note however what is being proposed: one conceptual distinction, speed, is determining that another, speed change, cannot be made directly, and by virtue of this ensuring the encapsulation of mechanisms and variables. In other words, a conceptual distinction is making the running as regards knowledge structure. On the theory-based approach, this would not be tolerated, for as emphasised from part I onwards the central tenet of the approach is that conceptual distinctions flow from mechanisms and not the reverse. However, in the context of the present chapter, this can hardly concern us. Nothing has emerged from our analysis of motion to provide encouragement for the theory-based perspective. From that perspective, children should treat the elucidation of mechanisms as a cognitive imperative, implying gradual improvement from the time mechanisms are recognised. Mechanisms relating to motion were recognised by the majority of the 6- to 7-year-olds in my 1991 study, but the age-related improvement alluded to above took place after 14. Prior to that, there was little sign of change with age, let alone improvement. Despite differing from the 1991 study over many results, the Twigger *et al.* (1994) research also locates age trends over mechanisms at a relatively late age. More importantly though, the motion research has not shown mechanisms to be the source of any conceptual distinctions, not speed certainly but not of course variables. Thus, just as the theory-based approach cannot accommodate the explanation for the encapsulation of variables and mechanisms which has been proposed, so it cannot accommodate the existence of encapsulation in the first place.

Accepting then that there is nothing in the present chapter for theory-oriented scholars, the next issue must be the implications for their action-grounded counterparts. Here, the situation is somewhat more promising. As stressed throughout, it is central to the action-based model that conceptual distinctions predate theoretical. Furthermore speed, the conceptual distinction of central concern, is exactly the kind of thing that the sensori-motion action stressed by the action-based perspective should give rise to. In addition, conceptual distinctions involving variables will not on the action-based model immediately be subsumed under mechanisms, even when the latter are known. From 11 onwards, children should experience some cognitive pressure to bring variables within the scope of mechanisms. However, as explained in the previous chapter, the putatively blocking effect of speed conceptions is entirely consistent with the action-based approach. Thus, in the particular case of motion, the continuing encapsulation poses no threat. All in all, then, the suspicion which emerged at the end of part II of this book has grown throughout part III: everyday physics requires the action-based and not the theory-based framework.

Acceptance of the action-based framework would undoubtedly take us forwards, but it would not resolve all the outstanding issues. Part II concluded by showing how neither the Piagetian nor the Vygotskyan version of action-based theorising could fully explain the heat transfer results. In the case of the Piagetian version, this was because heat transfer manifests theoretical structure from the early years of schooling, and scheme-language (strictly, scheme-symbol) co-ordinations as outlined by Vygotsky would be required to explain this. In the case of the Vygotskyan version, it was because understanding of the heat transfer mechanisms and their attendant variables shows a step-wise and not a steady improvement with age, and this implies that something additional to language is also at stake. Do the present results confirm these conclusions, and do they help us understand what an adequate action-based model would look like? To begin with the first question, the results are in truth neutral about the need for language. In the absence of theoretical structure, the co-ordinations relevant to motion could be between actions as Piaget proposed, or they could be between actions and atheoretical language. It is only the need for consistency with heat transfer that would incline us towards the latter.

As regards the need for something which mediates between language and the conceptual structures of children, this is confirmed by the present results. In the first place, the progress with variables and mechanisms was, if anything, more stepwise with motion than it was with heat transfer. As noted already, the improvements which occurred in my 1991 study took place around 14 and 15 when with heat transfer we were often talking about 11 or 12. Paralleling heat transfer, the problem is not so much that understanding was better at 14 or 15 than it was at, say, 12 or 13. This kind of progress could be explained in terms of formal teaching: the 14- to 15-year-olds were the only group in my 1991 sample to have studied Newton's Laws. Rather the problem is that the 12- to 13-year-olds were not substantially better than the 6- to 7-year-olds, despite having had six additional years of exposure to mature everyday physics as represented in language.

In addition though, the mechanisms which children generate for motion are not necessarily partial versions of mature everyday physics. For instance, earlier in the present chapter, we saw some children (typically in the 13 and under age group) proposing that air exerts a downwards force. Older children and presumably adults express the received view that air exerts an upthrust. To reiterate the point made above, young children must hear the views of older children and adults reflected in language, but if their own views are reversals this is further evidence of their being able to ignore what they hear. Moreover, given that their views are reversals and not downright ignorance, they must be able to bring some alternative information to bear. From the action-based perspective, the most likely source of this other information is the child's bodily experiences, and certainly the example given seems relatable to personal encounters with, say, wind. In addition of course, bodily experiences could explain the primary alternative mechanism reported in part II for heat transfer, transfer without transmission. Children do after all feel the fire's heat warming their bodies.

However, identifying the source of the alternative viewpoints does not explain why the linguistic messages are discounted in the first place. These messages must be processed. Otherwise, we could not explain why knowledge of heat transfer is structured theoretically from such a young age. However, something must happen to render the content of the messages unstable while preserving their structure. Could it be the feedback that children receive on their beliefs when *using* them in dialogue and action? Certainly, to the extent that this was negative (or seen to be negative by children), Inhelder and Piaget's (1958) vicious circle of directionless revision and further negative feedback would become a reality. Moreover, improvement after 11 or 12 would be predicted. However, if this is the line that we are going to take, we are converging on a story that should apply across the board, a story of appropriation from language, usage, feedback and loss, and gradual recovery a number of years later. Given the story, we can approach our third and final topic area with a sense of hypotheses to test. The topic area in question is object flotation, and in some respects it should be ideal for our purposes. It is true that, as part I explained, it has not been extensively researched for the preconceptions which students bring to physics instruction. Thus, we lack both evidence and claims concerning the endpoint of development. However what we have instead is a rich array of studies with younger children, and particularly the pre-teenage group which is emerging as crucial. For this reason, we can approach the studies with some anticipation of unearthing something useful.

Part IV Object flotation

7 Flotation in liquids and stage-like progression

So far, we have looked at two topic areas in physics, heat transfer and propelled motion. With heat transfer, we found that understanding of causal mechanisms showed a big improvement around 11 or 12, progress which was associated with a corresponding improvement in variable selection. As we probed further, the association between mechanisms and variables began to look like anything but an accident. On the contrary, there were strong indications of contingency, such that improvements in variables may have been dependent on improvements in mechanisms. As noted repeatedly, such contingency amounts to theoretical structure. With propelled motion, we found that understanding of causal mechanisms improved around 14 or 15, and that once again this was paralleled by progress over variables. However, this time, there was no evidence for interdependence. Rather, the indications were of two separate systems, one comprising the variables relevant to overall speed and the other the mechanisms by which speed change is 'perceived'.

With dependence on the one hand but total separation on the other, heat transfer and propelled motion seem on the face of it to be very different phenomena. Yet emerging at the end of part III was the sense that, despite the differences, development in the two topic areas could be the product of equivalent processes. In both cases, development may be traceable to preschool variables which are egocentric in nature. These variables are decentred around 6 through co-ordination with cultural symbols, primarily with language, and it is the nature of those symbols which determines the subsequent relation that variables will have with mechanisms. In particular, it is symbolic representation which dictates whether variables will depend on mechanisms as befits theoretical structure or whether the two will prove autonomous. Symbols may also provide information about the content of variables and mechanisms, but this information will be diluted or even lost through feedback on usage and the subsequent response to this. Conceptual constraints prevent systematic recovery from dilution much before 11, hence the absence of progress before that age in both topic areas.

Without doubt, much within the model is currently unclear. For instance, how exactly does decentration proceed, and how can usage qualify content while structure remains intact? Issues like these will have to be considered carefully, and they will in fact occupy much of part V. For now, we need further empirical evidence that the model is along the right lines, and this is what part IV will be concerned with. Recognising that the model is a hybrid of what has been referred to as the action-based approach, neither Piaget nor Vygotsky but something in-between, the question will be raised with reference to a new topic area of whether we really need to depart from these well-entrenched theories. In addition though, we need to be certain that the action-based approach is the correct one to follow. The alternative theory-based approach has had a rough ride so far, and particularly in part III. Nevertheless, we must proceed cautiously before rejecting it outright.

Noting the need to make further checks on the theory-based approach as well as our evolving model, there is everything to be gained by sticking with the strategy of parts II and III, that is a review of research relating to variables, mechanisms and their interrelation. This then, is how part IV will also be structured. However, given the emergent model, two pieces of evidence would be particularly welcome: (a) evidence for an early dependency of variables on mechanisms, that is early theoretical structure, because this would mean that if the action-based approach is correct decentration probably does depend on co-ordinations with symbols; and (b) evidence for delayed progress over the content of variables and mechanisms because this would mean first that the action-based approach probably is correct, second that symbolised information probably is lost in usage and third that cognitive constraints preventing early recovery probably do enter the equation. As it happens, part IV will be able to provide both kinds of evidence, but not interestingly in quite the form that appeared in part II.

The selected topic area is object flotation, an area which from the science point of view aligns closely with propelled motion. The situation where a speck of dust floats in the atmosphere is no different in principle from the situations where a parachute drifts slowly down to earth or a ball plummets from the top of a tower. The sole point of contrast is that in the first case the drag force is sufficiently great to permit a terminal velocity of zero before the ground is reached. The same is true of situations where objects fall through air but float after landing on a denser fluid like water, where for example a dropped beach ball floats on the sea. However while the point must be recognised, it is unlikely that more than a few individuals outside of professional science will be aware of the links. Indeed, even professional scientists did not see the links until relatively recently. The unitary perspective on flotation and motion is in fact a consequence of Newton's Laws, and these were only published a few hundred years ago. Prior to Newton's Laws, object flotation was treated by scientists as relatively self-contained, and attempts to explain it proceeded on this assumption.

As it happens, an adequate, albeit self-contained, explanation of flotation was proposed by the Greek scholar Archimedes over two thousand years ago. This explanation is usually referred to as 'Archimedes' Principle', and its central tenet is that when objects are immersed in fluids, either liquids like water or gases like air, they displace volumes of fluid equivalent to their own volumes. The displaced fluids exert an upthrust or 'buoyancy force' which depends on the fluids' weight. If the upthrust is greater than the downwards force exerted by the objects' weight, the objects will float. If it is less, they will sink. As a corollary, the variable that is crucial to predicting whether objects will float or sink is their density relative to the fluids they are immersed in, the formal definition of density being weight per unit volume. Ice is denser than alcohol but less dense than seawater. Thus, the upthrust from whisky displaced by ice cubes will be insufficient to counter the downwards force from the weight of the cubes, and the scotch will be truly *on* the rocks. The upthrust from seawater displaced by polar icebergs will on the other hand be more than adequate to counter the weight's downwards force, and the icebergs will accordingly float.

The existence of Archimedes' Principle as something self-contained, adequate and, it has to be admitted, relatively straightforward has proved consequential both for educational practice and for research. Within education, there has been a tendency to teach object flotation separately from Newton's Laws and at a somewhat earlier stage. Within research, object flotation has been sidelined from the debates regarding the everyday physics of propelled motion. Thus when as we saw in part III, researchers discuss whether the preconceptions of students entering physics do or do not manifest theoretical structure, they mean propelled motion excluding flotation. As noted in part I, no claims in this regard have been made about flotation. Yet as also noted previously, this does not mean that flotation has been ignored as a topic for research. On the contrary, there is a sizeable literature plotting children's variables and ascribed mechanisms against relative density and Archimedes' Principle. In the pages to follow, we shall review this literature, focusing in this chapter on flotation in liquids and in its successor on flotation in gases.

CHILD-BASED AND VARIABLE-BASED APPROACHES TO VARIABLE SELECTION

Children's experience with liquids is focused on water. However, with water, virtually all children encounter a rich array of instances of object flotation, including toys in the bath, household articles in the sink, sailing vessels on rivers, lakes and the sea, and their own bodies in the pool. As a result, questioning children about flotation in liquids typically makes sense to them, for they know that objects sometimes float and sometimes sink and they can see the point of trying to differentiate. On what however will differentiation be based? Do children have a sense of relative density, and if not what alternative variables are characteristically used? The present section will start with research that is relevant to the issue, showing that although the variables invoked for object flotation run into hundreds, relative density is seldom amongst them. Indeed relativism in any form is rare, with children concentrating instead on the absolute properties of objects and liquids.

Stages in variable selection

Although children differentiate floating from sinking, the point where they make the distinction is not necessarily consistent with the received wisdom of science. From the scientific perspective, objects are deemed to be floating if they are: (a) at the surface and partly immersed; (b) on the surface and not breaking surface tension; and (c) under the surface but freely suspended. A large-scale study by Biddulph (1983) suggests that only (a) is consistently used by children, and that even here other (scientifically irrelevant) conditions have sometimes to be fulfilled. Biddulph's study involved interviews with 104 pupils aged 7 to 14, and a written survey with 429 pupils aged 8 to 12. For both, pupils were shown line drawings where objects fulfilled or failed to fulfil conditions (a), (b) and (c), and asked if the objects were floating or not. Examples of the drawings include an apple one half above the water and one half under, a water spider standing on the surface, an underwater swimmer looking at fish, and a sunken bottle. Faced with such drawings, about 30 per cent of the overall sample denied floating when the objects were on top of the water (with denials being more frequent at the older age levels), and about 40 per cent denied floating when the objects were completely under. When the objects were partially immersed, some pupils required them to be stationary before they were seen as floating and some required them to be upright. For example, fifty-four pupils in the interview sample denied that an overturned yacht was floating.

Biddulph's results have a number of implications but in the present context the main one is methodological. If the intention is to establish the variables which children see as relevant to object flotation, the context of questioning must be one where definitional differences do not cloud the issue. Predictive problem solving should generally work here. If children are asked 'Will X float in this fishtank? And why?', divergence from the questioner over the semantics of floating should remain well hidden. Explanatory problem solving could also succeed, but to avoid confusion all presented instances would have to be paradigm cases, that is objects which are stationary, upright and partially immersed or well and truly sunken. The point is significant in view of the available literature, for several key studies focus on the explanation of observed events. Moreover, amongst these studies there are examples where paradigm cases have not been used. Rodrigues (1980) for example presented cartoons where the floating objects were almost always perched on the water's surface and where the sinking objects were often partially submerged. Piaget and Chatillon (1975) asked children to discuss the floating of completely submerged substances. The potential ambiguities here need to be borne in mind when considering the studies' results.

The work of Rodrigues and Piaget and Chatillon is in fact of relatively recent vintage, for research into children's conceptions of floating and sinking began as indicated back in part I with Piaget (1930). Piaget's work involved asking children aged 4 to around 10 whether (and why) small objects would float or sink in tapwater. The children's responses were noted down, and classified according to their adequacy. Four 'stages' were identified with reference to adequacy. During the first stage which was rare beyond 5 years, the variables relevant to object flotation were seen as personal and whimsical. Objects float when they want to or need to, with the implication that they can readily change their minds. During Piaget's second stage, variables that are physical were recognised for the first time, with objects being said to float because they are heavy. This stage, which was thought to be characteristic of 5- to 6-year-olds, was seen by Piaget as related to its predecessor: heaviness implies power and power can be personal. The third stage involved a reversal of the second for objects were now seen to float because they are light. More specifically, the small objects on which Piaget focused were expected to float when light and sink when heavy. When reminded of steamers on Lake Geneva, children giving third-stage responses who were typically 6 to 8 years of age would recognise the contradiction and call on special factors. Engines and movement, both relatable to second-stage 'power', were frequently used. During the fourth stage which was observed from 9 years onwards, children were believed to move from an absolute sense of lightness to the correct relativistic one: objects float when they are lighter than the same volume of liquid.

Piaget's 1930 work inspired a series of follow-ups of which the one reported by Laurendeau and Pinard (1962) is probably the most comprehensive. Some 500 children aged 4 to 12 were engaged in extended interviews. Interestingly, although the interviews followed Piaget in many respects, they focused on explanation rather than prediction. Children were asked to predict whether objects would float or sink but then, instead of justifying their predictions, they observed the outcomes and explained what happened. The explanations were used to determine stages of responding, of which there were once more four. The stages are not however exactly mappable onto Piaget's. In the first place, Laurendeau and Pinard's most primitive stage 'Stage 0' represented incomprehension, with its successor 'Stage 1' being equivalent to the personal, moral stage that Piaget reported. However consistent with Piaget, neither Stage 0 responding nor Stage 1 was frequent beyond the age of 5.

Laurendeau and Pinard's 'Stage 2' constitutes a further point of contrast with Piaget. This stage was defined by the usage of exclusively physical variables and was characteristic of over 50 per cent of the sample between the ages of 6 and 10. It comprised two substages which can be related in some respects to Piaget's second and third stages. Nevertheless, there is a difference because Piaget's stages referred exclusively to weight while Laurendeau and Pinard's substages included shape, substance and size (the difference between the substages being in how systematically the variables were used). Because systematicity was important within Stage 2, the distinction between this stage and Laurendeau and Pinard's final Stage 3 is not entirely clear. The first substage of Stage 3 also involved variables like weight, shape, substance and size but now usage was entirely systematic. Creeping in, though, was some relativism, for example 'lightness for size', 'lightness for the water'. However, unlike Piaget, Laurendeau and Pinard were careful to differentiate such vague relativism from the more precise concept of relative density. This may be why just two of their 500 children were deemed to have reached the final substage of their Stage 3, which centred on relative density.

Piaget wrote a preface to Laurendeau and Pinard's report, a preface in which he commented only on the similarities with his 1930 work. The differences were glossed over. This is odd in that Piaget had at that time just returned to object flotation via research summarised by Inhelder and Piaget (1958). His representation of the new data was, if anything, even more divergent from Laurendeau and Pinard than the 1930 effort. The Inhelder and Piaget work was based on a study where 4- to 14-year-old pupils were asked to classify small objects into those that would float in a bucket of water and those that would sink. Once all the objects had been classified, the pupils were asked to justify their selections, justifications being placed into one of three stages.

During the first of Inhelder and Piaget's stages which was said to occur up to 7 or 8 years, the pupils progressed from being unable even to classify to classifying but being unable to generate a unifying principle. Thus, one 5-and-a-half-year-old designated some objects as floaters because of their nature, for example ducks, others because of their smallness, others because of their lightness, others because of their flatness and a further group because of their colour. Such a pupil would be flummoxed by the fact that large light objects are more likely to float than small light objects, and it is this contradiction which according to Inhelder and Piaget triggers the second stage. During this stage which lasts from 7 to (perhaps) 10 years of age, pupils were said to wrestle with the contradiction and through this to shift to a relativistic sense of weight. Objects float when they are 'made of light stuff' or 'light enough for the water'. However, the inaccessibility, as Inhelder and Piaget saw it, of the concept of volume is believed to preclude entry to the third stage until 11 or 12. Like the final substage of Laurendeau and Pinard's Stage 3, the third stage is characterised by true appreciation of relative density.

Inhelder and Piaget's second and third stages are in fact equivalent to the two substages of Laurendeau and Pinard's final stage. However, this means that their first stage incorporates all of the latter's other stages, as indeed it incorporates the first three stages of Piaget (1930). It is not clear why Inhelder and Piaget did this. Was it because responses falling into one or more of the lower level stages were too rare to bother with? Was it because the cognitive processes implied by these stages were insufficiently interesting to merit attention? The second possibility would be consistent with the overall tenor of Inhelder and Piaget's book, for as we noted in part I this is where they discuss the final stage of action–entity differentiation, the possibilistic stage. This is indeed where they introduce the term 'formal operations', the best known of the several labels which they use for the stage.

However, if cognitive processes were the rationale for Inhelder and Piaget's approach they cannot have been thought through carefully, for it would be hard to find a general cognitive account which motivated Inhelder and Piaget's second stage as a direct successor of their first. Although defined in terms of the absence of unifying principles, the first stage was assumed also to involve non-unifying principles which lead to contradiction. The second stage is entered when a unifying principle is discovered which avoids contradiction. However why should both problems be solved simultaneously? Is it not possible that some pupils derive unifying but contradictory principles or the reverse? Is this not particularly the case if we construe the problem in general cognitive terms? Rodrigues (1980) is one researcher who seems to recognise the point for she reported what amounts to 'unification with contradiction' in three pupils (from a sample of fifteen) whose average age was just over 9. As mentioned earlier, Rodrigues' methodology is open to criticism. Nevertheless, it is interesting that she recognised the tendency in what she called a 'Type (b) response, a response which is securely based on a single variable' (Rodrigues, 1980: 65). This is particularly the case when her Type (a), (c) and (d) responses were chronologically and conceptually relatable to Inhelder and Piaget's stages.

Table 7.1 schematises the relations between the work of Rodrigues and Inhelder and Piaget. To complete the picture, it also shows the relations between these pieces of work and the other two discussed in the section. It can be inferred from Table 7.1 that even if we discount Laurendeau and Pinard's Stage 0, five levels would be needed to cover all of the distinctions which the theorists are making. These levels are: I. failure to use physical variables, e.g. personalism; II. use of irrelevant physical variables, for example flatness, temperature and colour (as aids to floating or sinking) and heaviness or heaviness relative to the water (as aids to floating); III. use of relevant variables in an absolute sense, for example largeness and lightness (as aids to floating); IV. relativistic use of relevant variables, for example lightness for size, lightness for the water; V. relative density. None of the theorists recognise all of these levels, and if the principles which Howe (1991) applied with heat transfer and propelled motion are carried through to object flotation, I cannot do this either. Indeed, as will be recalled from earlier chapters, my 1991 study made do with what were in effect two levels when dealing with its heat transfer and propelled motion data: relevance (equivalent to levels III, IV and V and irrelevance (equivalent to levels I and II).

However, while my 1991 scheme is relatively undiscriminating in terms of level, it is arguably more sensitive in other respects. It was used to represent *variables* while the schemes in Table 7.1 were used without exception to

Table 7.1 Summary of four stage-based approaches to flotation in liquids

Piaget (1930)	Laurendeau & Pinard (1962)	Inhelder & Piaget (1958)	Rodrigues (1980)
	Stage 0		
Stage 1	Stage 1		
Stage 2		Stage 1	Type (a)
	Stage 2		
Stage 3			Type (b)
		Stage 2	Type (c)
Stage 4	Stage 3		
		Stage 3	Type (d)

represent *children*. Based on the variables they proposed taken as a totality, children were assigned to particular stages. Variable- and child-based approaches may lead to equivalent conclusions if children are consistent (or nearly consistent) in the levels of their variables. However, the data presented earlier for heat transfer and propelled motion must make this seem unlikely, and there are in fact instances of marked inconsistency for object flotation throughout the work discussed in this section. For example, Inhelder and Piaget quote one participant as saying 'Ah! They (first objects) are too heavy for the water so the water can't carry them . . . . They (second objects) are quite light . . . . They (third objects) can go to the bottom because the water can go over the top' (Inhelder and Piaget, 1958: 32). Here we have variables that would be coded at Levels IV, III and II respectively in the scheme just presented. In view of such instances, circumspection would seem in order, with the implication that first pass coding schemes should operate at the level of variables. By the same token, though, there might be mileage for now in suspending the two-level approach of Howe (1991), revealing though this may have been to date, and exploring whether the five levels signalled by Table 7.1 have anything to add.

Variable-based approaches to variable selection

Having endorsed a variable-based approach in principle, we immediately run into a problem when trying to implement it in practice: exemplars of its use with object flotation are few and far between. Indeed, the work of Biddulph (1983) that was introduced earlier stands virtually alone amongst published research. Biddulph, it will be remembered, used one-to-one interviews and a written survey. The interviews, but not the surveys, included items where ten simple objects, for example a wooden block, a school eraser, a plasticine ball and a piece of candle were immersed in water. Subsequent to immersion, the pupils were asked to explain why the objects floated or sank and the explanations were then coded for the variables mentioned. The variables most commonly associated with floating were being light, having air inside, being of buoyant material and being non-compacted. The variables most commonly associated with sinking were being heavy, having no air inside, not being of buoyant material, being solid and being absorbent. Interestingly, 49 per cent of the explanations for floating and 44 per cent of the explanations for sinking involved more than one variable. Thus, we are talking about variable clusters and not isolated variables, raising the question of whether the clusters chosen by given pupils were homogeneous as regards level of understanding. Unfortunately, we cannot tell from Biddulph's data because his coding of variables ignored appropriateness and the information is not retrievable from what he presented.

Thus, while the studies discussed earlier ignored variables in the interests of level, Biddulph ignored level in the interests of variables. We need research which allows us to take variables as the unit of analysis while simultaneously considering level. A preliminary attempt to do this was reported by Stepans *et al.* (1986). Using the prediction and justification-of-prediction method rather than Biddulph's observation and explanation, Stepans *et al.* interviewed 184 students about the flotation of twelve small objects. The students varied from kindergarten to college level, with ages presumably ranging from 5 to 18 and over. For purposes of analysis, students were divided into four age groups, but in the event this proved unnecessary for there were no significant age trends. About one-quarter of the sample gave responses which showed 'no understanding' of the variables relevant to flotation,[1] about three-quarters gave responses which showed 'partial understanding', and a miniscule proportion (ranging from 0 per cent of the youngest to 2 per cent of the eldest) gave responses which showed 'complete understanding'. Complete understanding is, we can assume, equivalent to relative density, labelled 'Level V' in the scheme outlined earlier.

The rarity of complete understanding is probably the most striking aspect of Stepans *et al.*'s results, for it flies completely in the face of Piagetian claims; see both Piaget (1930) and Inhelder and Piaget (1958). However, there is a second and perhaps less obvious point to be made which relates to the sense that can be gained from the partial understanding, the majority response. If we refer back to the earlier definition of Levels II to IV, a case can be made for each one showing partial understanding when compared with the level below. Indeed, even Level I shows partial understanding when compared with Laurendeau and Pinard's (1962) 'Stage 0' of incomprehension. Since Stepans *et al.* do not define their 'partial' category further, it is impossible to know what it maps onto. This is unfortunate because if, say, partial understanding corresponded with my 1991 notion of relevant variables and no understanding with my notion of irrelevant, the absence of age trends would set object flotation apart from both heat transfer and propelled motion. In the context of this book, it would as a consequence be of some theoretical significance.

Thus, Biddulph's study by-passes levels and Stepans *et al.* uses them in a fashion which in the present context is rather unhelpful. Since these studies are more-or-less the sum total of published variable-based research, we are thrown once more on unpublished material, with this time two sets of data to call upon. The first set derives from a study by Howe *et al.* (1990) which had a somewhat different agenda from the one we are following here, and the second comes once more from the investigation of Howe (1991) which of course had the present agenda as its focal concern. The Howe *et al.* study involved 121 pupils from a Glasgow primary school, twenty-seven aged 8 to 9, thirty-two aged 9 to 10, thirty aged 10 to 11, and thirty-two aged 11 to 12. The pupils were taken one-by-one through a four-part interview. During Part I, they were presented with eight small objects, these including a glass bottle, a wooden ball, a plastic button and a metal dish. As each object was presented, the pupils were asked whether it would float or sink in a tank of tapwater, and why they thought this. Once all the objects had been discussed, the pupils were invited to try them in the tank and place them in boxes labelled 'Floaters' and 'Sinkers' according to what had happened. During Part II, eight real-world examples were described to the pupils, for example boats floating in the sea, coins sinking in wishing wells, people floating in a swimming pool and potatoes sinking in a saucepan. The pupils were asked to explain why the examples turned out as they did. During Part III, attention was re-directed at the eight small objects and the pupils were asked to imagine them immersed first in water in other locations (Loch Lomond, a mountain stream, a half-filled saucepan) and second in the tank with other liquids (whisky, milk). Questioning was directed at whether (and why) the 'Floaters' would still float and the 'Sinkers' still sink. Finally during Part IV, five real-world examples were described which related to the effects of liquids, for example the Dead Sea vs. other water and orange juice vs. whisky. The pupils were asked to explain why the liquids had their effects.

The methodology obviously has similarities with the studies already discussed, but there are two points of contrast. First, the pupils were questioned more extensively than is typically the case. In other words, they were not simply asked 'Why does/will X float?' but also whether floating would always occur in the presence of whichever variables the pupils mentioned. Likewise for sinking. Second, by virtue of Parts III and IV, the pupils were provided with contexts where liquids were highlighted. Insofar as flotation is determined by the densities of objects relative to liquids, it is the object–liquid relation that children should be focusing on. Some children clearly do this but all the studies discussed earlier indicate that many do not, with the focus being on absolute properties rather than relations. However, these absolute properties seem overwhelmingly to relate to objects (see, for example, the list associated with Biddulph, 1983). Does this mean that absolute properties of liquids are treated as irrelevant, or could it be that the context of questioning focuses attention on objects? Two considerations made Howe *et al.* suspect that it might be the latter.

First, it was that the contexts of questioning typically involved variability in objects but constancy in liquids (tanks of water). Second, the occasional reports in the literature of references to liquids indicated that liquid properties may not be entirely discounted. For example, in accounting for floating, one of Inhelder and Piaget's (1958) interviewees stated 'It's too big and then there's too much water' (Inhelder and Piaget, 1958: 26) and one of Laurendeau and Pinard's (1962) claimed 'There is quite enough water' (Laurendeau and Pinard, 1962: 223). Biddulph (1983) asked his sample whether the depth of the water would affect the level at which boats floated. A total of 25 per cent said boats would float at a lower level in deep water; 16 per cent said they would float at a higher level; and the remainder said that the depth would make no difference. Finally, Piaget and Chatillon (1975) presented pupils aged 4 to 12 with a vessel containing liquids of different densities. The pupils were asked to explain why drops of oil fell through one liquid but stayed suspended in another. The main result was a shift with age from object-based explanations to liquid-based ones.

Considerations such as the above persuaded Howe *et al.* (1990) to design an interview schedule where both object variables and liquid were equally highlighted, hence Parts III and IV as well as Parts I and II. Analysis of the responses showed that liquid variables were indeed mentioned. However, the most striking feature of the data was not this but rather the overall quantity of variables, both object and liquid that the pupils chose to cite. Outstripping anything reported in the literature, they mentioned 135 different variables which they perceived as relevant to floating and 115 which they perceived as relevant to sinking. Some 73 per cent of the variables relevant to floating were properties of objects, 23 per cent were properties of liquids and 4 per cent were properties of the object–liquid relation. The equivalent percentages for variables relevant to sinking were 70 per cent, 27 per cent and 3 per cent respectively.

As Table 7.2 shows, a small number of the variables which the pupils mentioned were codable at Level I indicating that non-physical properties were occasionally used. These properties included the bravery, concentration and effort of objects, and the goodness, horribleness and protectiveness of liquids, thus 'Things float in the Dead Sea because no-one wants to go under' and 'Ice floats in orange juice because orange juice is good, but it sinks in whisky because whisky is bad' (both produced by the same 8-year-old) and 'People float because they're brave and patient' (produced on three occasions, by two 8-year-olds and one 10-year-old).

Table 7.2   Absolute and relational variables as a function of level: flotation in liquids (Howe *et al.*, 1990)

	Object variables relevant to floating	Fluid variables relevant to floating	Object–fluid relations relevant to floating	Object variables relevant to sinking	Fluid variables relevant to sinking	Object–fluid relations relevant to sinking
I	5	2	0	2	1	0
II	88	25	3	74	26	1
III	5	4	N/A	3	4	N/A
IV	0	0	3	1	0	3
V	N/A	N/A	0	N/A	N/A	0
TOTAL	98	31	6	80	31	4

However, while Level I variables were used they were nothing like so numerous as Level II, the level of non-relevant physical properties. At Level II, there were irrelevant references to weight, objects floating when heavy or middle weight and liquids supporting floating when light or light relative to the objects. Particularly intriguing were the views of two 9-year-olds and one 10-year-old that 'Light things float in light liquids and heavy things in heavy liquids'. Similarly, there were irrelevant references to size, small objects floating and large ones sinking. However, far outstripping these 'reversals of relevancy' were variables that were irrelevant no matter how they were used. These included variables which called upon: (a) shape, for example: round, pointed, flat, arrow-shaped, not straight, log-shaped, air-holding shape; (b) orientation, for example: tilted, leaning back, balanced on water, stuck to the top; (c) surface, for example: smooth, prickly, painted, muddy, peeled, without holes; (d) constitution, for example: metal, cork, wood, alcoholic, pasteurised; (e) contents, for example: hollow, air, blood, water, sugar, calcium, vitamins, germs; (f) parts, for example: propellers, engines, armbands, stones, bellies, dinner and 'special holding up things'; (g) temperature, that is heat and cold of both objects and liquids; and (h) movement, that is motion and stillness again of both objects and liquids. There were also idiosyncratic notions that defy grouping, for example 'Glass can't be seen so you can't push it down' and 'Icebergs float because they've been here a long time' (produced by an 8- and a 9-year-old respectively).

By their very nature of involving relevant properties only, Levels III, IV and V offer less scope for variability, and this is reflected in Table 7.2. Nevertheless, five Level III object properties, big, thin, spread out, light and light interior, were offered for floating and three, small, thick and heavy, were offered for sinking. Four Level III liquid properties, salty, strong, thick and heavy, were offered for floating and four, without salt, weak, thin and light, were offered for sinking. At Level IV, it was striking that most of the variables related objects to liquids, thus 'Orange juice is thicker than ice, so the cubes will float' and 'Metal is stronger than water, so the spanner will sink'. Relations between relevant object variables were rare, amounting merely to 'heaviness for size' which was used once by each of three pupils all aged 12. As for Level V, the story can be quickly told: there were no mentions whatsoever of variables codable as relative density.

How then were the variables distributed across pupils? Did each pupil use a single variable or did some use a cluster? If the latter, was there consistency in the variables that co-occurred and was the consistency also with reference to level? In analysing the data with these questions in mind, I decided to integrate the variables associated with floating with the variables associated with sinking, since the former were often (as earlier examples will have shown) reversals of the latter. Thus, pupils who said that light objects float and heavy objects sink were regarded as using a single variable. Integration was of course only allowed when the variables associated with floating and sinking were true reversals: pupils who said that light objects float and light objects sink or that light objects float and heavy objects float were deemed to use two variables. As it happened, integration did nothing to eclipse the enormous heterogeneity across the sample as a whole and indeed within individual pupils. No pupil used fewer than three variables, and two used over thirty. Moreover, as Figure 7.1 shows, nearly 80 per cent of the sample mentioned at least ten variables.

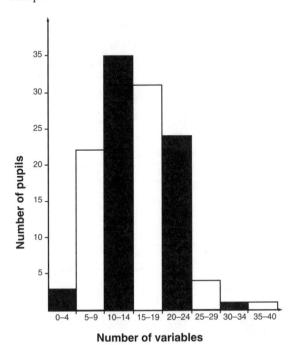

Figure 7.1 Distribution of pupils by number of variables used

Source: Howe *et al.* 1990.

Were there any patterns to the variables used, in that use of one variable was consistently associated with use of others? A first pass through the data was not encouraging. Of the 209 variables that remained when the variables associated with floating and sinking were combined, 42 per cent were used by one pupil only, obviating any patterns to their usage. Nevertheless some variables were used with sufficient frequency to recommend a formal cluster analysis and this was carried out.[2] The results were revealing in their negativity, for no robust clusters whatsoever were identified in the sample. Perhaps, though, pupils were systematic over level, choosing variables from one level at the expense of the others. If that were the case, there should be negative correlations across the sample between the number of variables at each level. To explore this, two correlation matrices were computed. The first involved the number of distinct variables used at each level by each pupil regardless of number of mentions. Thus, a pupil who referred only to the presence or absence of air and the presence or absence of wood would be credited with two Level II variables and no others. The second matrix involved the number of times variables at each level were mentioned. Thus, if our hypothetical pupil referred to air on six occasions and wood on four, she would be credited with ten Level II mentions. The matrices are presented in Table 7.3 where it is clear that positive correlations are both more numerous than negative ones and larger in magnitude. Indeed, all statistically significant correlations in Table 7.3 are positive.

It appears then that the clusters of variables used by 8- to 12-year-olds are both divergent across pupils and mixed with regard to level. This raises innumerable questions about the studies with which the section began but the foremost must be surely where does this leave us with regard to the claims they make about changes with age? As a first step to finding out,

Table 7.3   Correlations between frequencies of usage at different levels: flotation in liquids (Howe *et al.*, 1990)

	No. of distinct variables		
	Level I	*Level II*	*Level III*
Level II	+0.20*	-	-
Level III	-0.15	+0.25**	-
Level IV	+0.05	+0.05	+0.17

	No. of variable mentions		
	Level I	*Level II*	*Level III*
Level II	+0.25**	-	-
Level III	-0.02	-0.01	-
Level IV	+0.03	-10.03	+0.02

* Indicates statistical significance at p<0.05
**Indicates statistical significance at p<0.01

Table 7.4 shows the mean number of distinct Level I, II, III and IV variables produced by the four age groups featured in the study. It also shows the mean number of mentions of Level I, II, III and IV variables.

From Table 7.4 it is clear that neither Level I variables nor Level IV were used with any frequency. There were few instances of variables at these levels, and in most cases few mentions of the variables that were occasionally used. The only exception to the latter was with Level IV variables in the 10- to 11-year age group. The unusually large standard deviation reflects the fact that one pupil talked about being 'lighter than the water' on sixteen occasions while another did this on ten. Despite this, one-way ANOVAs produced no significant age differences for Level I and IV, not for distinct variables nor for variable mentions.

With Levels II and III, the picture is rather different. In the first place, usage of these variables was far from infrequent, particularly at Level II. In the second, there are indications in Table 7.4 of age differences in usage, with both Level II and Level III variables being more frequent in the oldest age group than they were in the youngest. This said, one-way ANOVAs on the Level II data showed the age differences to fall some way short of statistical significance (F for distinct variables = 1.72, df = 3,120, p<0.20 and F for variable mentions = 1.58, df = 3,120, p = 0.20). It was only with the Level III data that statistical significance was obtained (F for distinct variables = 4.85, df = 3,120, p<0.01 and F for variable mentions = 2.67, df = 3,120, p=0.05). Follow-up Scheffé tests showed that with both analyses, the 11- to 12-year-olds outstripped the other age groups, who did not differ amongst themselves. Interestingly, this difference was completely explicable

Table 7.4   Usage of variable levels as a function of age: flotation in liquids
(Howe *et al.*, 1990)

| | No. of distinct variables | | | | | | | |
| | *Level I* | | *Level II* | | *Level III* | | *Level IV* | |
	M	SD	M	SD	M	SD	M	SD
8–9-year-olds	0.44	1.22	10.00	5.87	1.67	0.78	0.48	0.64
9–10-year-olds	0.28	0.58	12.91	6.65	1.84	1.02	0.47	0.62
10–11-year-olds	0.43	0.73	12.13	5.10	1.87	0.90	0.53	0.68
11–12-year-olds	0.06	0.25	13.00	4.83	2.59	1.32	0.50	0.76

| | No. of variable mentions | | | | | | | |
| | *Level I* | | *Level II* | | *Level III* | | *Level IV* | |
	M	SD	M	SD	M	SD	M	SD
8–9-year-olds	0.48	1.40	17.04	10.91	10.74	5.13	0.89	1.40
9–10-year-olds	0.34	0.75	21.28	10.95	9.69	4.07	0.97	1.69
10–11-year-olds	0.50	1.01	20.27	8.73	10.00	3.91	1.73	3.47
11–12-year-olds	0.06	0.25	22.47	9.09	12.53	4.47	1.22	2.24

(Level V is excluded because of non-usage)

in terms of Level III liquid properties, with no differences between the 11- to 12-year-olds and the other pupils over Level III object properties. With the combining of the variables relevant to floating and the variables relevant to sinking, there were four Level III liquid variables: saltiness, thickness, strength and heaviness as facilitative of floating. The 11- to 12-year-olds averaged 1.41 of these variables, the younger pupils 0.78 (t = 2.32, df = 119, p<0.05). There were correspondingly five Level III object variables: light-ness, bigness, thinness, being spread out and having a light interior as facilitative of floating. Here the 11- to 12-year-olds averaged 1.18 and the younger pupils 1.01, a non-significant difference.

Reviewing the Howe *et al.* results, there is a key methodological point to be made which stands apart somewhat from any theoretical implications that need to be drawn. Specifically, although it was probably worth using the five levels in the interests of thoroughness, they did not turn out to be needed in practice to represent the children's knowledge. Levels I, IV and V were used rarely if they were used at all, and all the interesting age trends occurred between Levels II and III. This means that since Level II and below vs. Level III and above is the cut-off between irrelevancy and rele-vancy, the seemingly cavalier approach of Howe (1991) has received some vindication. Certainly, the vindication was sufficient to continue the approach when analysing the subset of the 1991 data which itself related to flotation.

The subset was obtained in the 1991 study to check the finding of the Howe *et al.* study that understanding of relevant liquid properties increases with age. To this end, the 126 pupils who featured in the 1991 study were interviewed about two photographs, one depicting an airbed on the verge of sinking and the other a basket of aquatic plants which was bobbing around in a pond. The pupils were initially asked in a general way whether there was anything that could be done to the water to make the airbed float and the plants sink. Depending on what they said, some, all or none of a set of more specific questions followed. These related to the usefulness of changing the roughness, depth, temperature and constitution of the water. Responses were noted and whenever a variable was proposed the pupils were asked why this was relevant. Although these variables were subsequently coded as irrel-evant or relevant, as signalled above, it is worth noting that all but one would have been coded at Level II or Level III in the more discriminating scheme which evolved from Table 7.1. The exception referred to changing the water's density, and was used by eight pupils with the airbed scene and two with the plants. These pupils were all in the 12- to 13-year-old age group or the 14- to 15-year-old.

For each scene, a count was made of the number of distinct relevant vari-ables and the number of distinct irrelevant variables with means being computed for each of the age groups. These means are shown in Table 7.5, and it should be noted that because they are presented on a scene-by-scene basis 'variable mentions' in the sense of Tables 7.3 and 7.4 are derivable from them. The main point to note from Table 7.5 is that with the airbed scene there was an age-related increase in the usage of relevant variables. This increase was statistically significant ($F = 3.82$, df $= 4,121$, $p<0.01$), with follow-up Scheffé tests showing the two youngest age groups to have produced significantly fewer relevancies than the three oldest. Interestingly, an equivalent result was not obtained for the plants scene, where the extreme rarity of relevant variables precluded any age differences whatsoever. Perhaps water is the thinnest, purest, least dense liquid that most pupils could spontaneously contemplate. Thus, when asked what they could do to make the plant basket sink, even those who thought along the right lines were flummoxed. This said, it is clear from the data for both scenes that even when the pupils were on the right lines in terms of relevancies, they must have still persevered with irrelevancies for irrelevant variables outnumbered relevant at all ages studied. However, while this was true, irrelevant variables showed a curvilinear relation with age, a relation which produced statistically significant age trends with the plants scene ($F = 2.62$, df $= 4,121$, $p<0.05$) although not with the airbed.

The important point is of course that the major division over relevancies and irrelevancies occurred around 10 and 11. Relevancies started to increase then, and irrelevancies peaked to decline thereafter. Thus, improvement at the level of variables was a late phenomenon, exactly as predicted by our emergent model. Moreover, the lateness was not specific to the 1991 study for it was also observed with the Howe *et al.* (1990) data. The only slight discrepancy was that with the latter, the significant differences were between the 11- to 12-year olds on the one hand and the 8- to 11-year-olds on the other. As with the slight discrepancy observed in relation to heat transfer in part II, this could reflect differences relating to demography and/or rapport. The Howe *et al.* sample varied from lower middle class to acutely deprived while, as mentioned already, the 1991 pupils came from an extremely affluent part of the country. In addition, the Howe *et al.* sample had never previously met their interviewer while there was a high level of personal acquaintance between the 1991 interviewer and her subjects. Whatever its

Table 7.5 Mean number of relevant and irrelevant variables as a function of age: flotation in liquids (Howe, 1991)

	6–7-year-olds	8–9-year-olds	10–11-year-olds	12–13-year-olds	14–15-year-olds
Airbed scene					
Relevant	0.00	0.12	0.29	0.46	0.41
Irrelevant	1.12	1.00	1.29	1.13	1.04
Plants scene					
Relevant	0.08	0.08	0.00	0.00	0.07
Irrelevant	0.65	0.60	1.08	0.46	0.74

source, the discrepancy does nothing to detract from the similarity between the two sets of data, a similarity which seems extremely important given our guiding issues.

Child- and variable-based approaches compared

The similarity between the results of my two studies could be important. Nevertheless, it must not be forgotten that the results are in marked contrast with the four pieces of work introduced at the start of the section, Piaget (1930), Laurendeau and Pinard (1962), Inhelder and Piaget (1958) and Rodrigues (1980). There, the message was that most children do not simply rid their knowledge systems of irrelevancies by 9 or 10 but also attain relevant variables which are equivalent to Level IV (if not Level V). In my studies irrelevancies continued up to 14 and 15, and relevancies in any form were rare until 10, 11 and 12. Moreover, the bulk of the relevancies were Level III, with the age-related growth being mainly in terms of Level III numbers.

The contrast between the two sets of results is unlikely to reflect the data-gathering procedures. The procedures were broadly similar in all six cases, with mine if anything being more searching. Likewise participant sampling is also unlikely to be responsible, given what was said above about how my two studies differed between themselves. We seem to be left with the approach to analysis, variable-based in my case and child-based in the others, and certainly the two other variable-based studies mentioned earlier, Biddulph (1983) and Stepans *et al.* (1986), point to conclusions which are more similar to mine than to the child-based research. Biddulph, it will be recalled, found pupils emphasising weight, airiness, material and solidity, all codable at Levels II and III. Interestingly, Biddulph also notes explicitly that very few pupils mentioned lightness for size. Stepans *et al.* found very few students showing complete understanding of relative density, even at college age.

Faced with having to choose between two sets of results, it seems on the face of it as if we must follow the variable-based message. Just as the child-based approach was seen earlier to ignore inconsistencies, so it must run the risk of overly optimistic syntheses. The plodding variable-by-variable approach carries no corresponding danger of underinterpretation. If it was not for a recent study by Kohn (1993), I should at this point rest my case and reassert the message from my own two studies. However, mention must be made of Kohn because her work could be read as suggesting that even Piaget and co. have underestimated children's knowledge!

Kohn's work used procedures which contrast in some respects with all the studies discussed so far. These procedures centred on two sets of eighteen blocks. The blocks in both sets were made out of wood, but with one set the blocks were covered with aluminium sheeting to make them look like metal. Within each set, the blocks were systematically varied along the dimensions of volume, weight and density, but they were placed in a random order for presentation to the participants. The participants were twenty college students, fourteen children aged 3 to 4 and fourteen children aged 4 to (nearly) 6. The task for all participants was to handle each object in turn, and predict whether they would float or sink in a basin of water. Some 86 per cent of the college students, 72 per cent of the 4- to 6-year-olds, and 53 per cent of the 3- to 4-year-olds predicted in accordance with density. Kohn would, I think, take this as indicative of Level IV responding, if not Level V.

The point is that if children of 4 and under can engage in Level IV responding, all the material presented in this section would appear to be in doubt. However, is this the correct gloss to place on Kohn's otherwise interesting data? I think not. What Kohn is tapping is a kinesthetic awareness of density, and it is of course important to find it established so young. However, kinesthetic awareness is not the same as understanding of variables. If it was, boat building could never have taken place until Archimedes had discovered his Principle! Kohn's data would have been somewhat more compelling if her sample had been prevented from handling the objects. However, even then there would have been problems, remembering the ambiguities with exclusively predictive responses that were discussed in part I. Noting this, I feel that we are stuck with the predict/observe and explain methodologies that have been a feature of the section. As a result, we are left with a choice between the relatively rapid development reported by the child-based investigations and the relatively slow development reported by their variable-based alternatives. For reasons expressed already, I think we have to go with the latter.

As noted earlier, the relatively slow development is of potential theoretical importance for it concurs with our emergent model. However, to appraise its importance fully, we need to know something more about the knowledge structures in which variables are embedded, and in particular about the relations that variables have with mechanisms. In this context, another feature of the data might appear to gain in significance, namely the heterogeneity between children over the variables chosen. Over 200 distinct variables were generated by the Howe *et al.* (1990) sample, with nearly half of these being unique to specific children. Quite apart from being once more at variance with earlier research (particularly the contributions made by Piaget), this heterogeneity does not seem to augur well for a significant role for mechanisms in object flotation. The point about mechanisms is that they are supposed to constrain the selection of variables, but if variable selection is not constrained in the first place the possibility seems somewhat beside the point. Somewhat but not necessarily entirely. As mentioned earlier, the heterogeneity was not completely unbounded, for some variables (albeit a minority) recurred in the Howe *et al.* data. Perhaps these variables were motivated by mechanisms. Perhaps also mechanisms played some part in the greater use of relevant variables from 10, 11 or 12. Recognising these points, there is still some purpose in asking our familiar questions: (a) do children call on mechanisms with flotation in liquids; and (b) do these mechanisms bear in a generative fashion on variable selection? These questions will be addressed in the following section.

THEORIES IN THE OBJECT–LIQUID INTERPLAY

As signalled earlier, an adequate account of the mechanisms by which flotation in liquids is achieved has existed for a very long time. This account centres on the liquid displaced when objects are immersed, liquid which creates an upthrust or buoyancy force in opposition to the downwards force from the object's weight. Statable in a single sentence, the account appears quite simple. Thus, there seems some chance first that children subscribe to mechanisms and second that these mechanisms resemble science 'truths'. Is this what happens in practice? The present section will start by attempting to find out. It will conclude that although mechanisms are abundant in the thinking of children, they can be way off beam when it comes to science orthodoxy. This will prove important when the section turns to the generativity of mechanisms in relation to variables. Its key claims here will be that first, mechanisms seem to motivate some variables but by no means all, and second, this partially as opposed to fully theoretical structure is itself a by-product of mechanism unorthodoxy.

Downthrust, barriers and upthrust in liquids

As it happens the published research is not particularly forthcoming with regard to mechanisms. Of the studies discussed in the previous section, Laurendeau and Pinard (1962), Inhelder and Piaget (1958), Rodrigues (1980) and Stepans *et al.* (1986), none of them make any mention of mechanistic knowledge. In the case of Inhelder and Piaget and Rodrigues, there is an explicit restriction to 'law-like' thinking. Piaget (1930) is centrally concerned with the relation between laws and theories, yet his chapter on object flotation gives the latter the most cursory of mentions. All that is said is that when during the third stage children claim that big things float they mean that bigness implies a bigger force from water. However, no evidence is provided to support this proposal, neither from the children's comments nor from what they imply.

Biddulph (1983) is somewhat more helpful, for his interview sample of 144 7- to 14-year-olds was asked about displacement. In particular, the pupils were shown a hollow box made from plasticine and asked first whether it would 'push out' the same amount of water as a plasticine ball of equivalent volume and second whether it would push out the same amount of water when floating as it would when submerged. Only about 33 per cent of the sample gave the correct 'same amount' answer to the first question, with correctness being more frequent in the younger pupils than it was in the older. With the second question, correctness was at around the 25 per cent level, this time without age proving a significant factor. These results are revealing in that, without a firm grasp of displacement, children stand no chance of adequate mechanisms in the context of flotation. However, ambiguity remains on two scores. First, it is not clear whether the pupils whose understanding of displacement was limited subscribed to alternative mechanisms for flotation or to no mechanisms at all. Second, it is also unclear whether the pupils who grasped displacement understood the significance of this in the process of floating. Did they, for example, realise that the pushed out water exerts an upthrust? Did they recognise this upthrust as the key causal mechanism in countering the objects' weight?

To redress these problems, the Howe *et al.* (1990) study used the more elliptical approach to eliciting mechanisms that figured (and was justified) in previous chapters, namely following 'Why does/will X float/sink?' with 'Why is (variable) important to floating and sinking?' Analysis of the responses that the pupils gave revealed that in twenty-three cases the question 'Why is (variable) important . . . ?' had consistently received the answer 'Don't know'. In other words, twenty-three pupils had said 'Don't know' on each of the twenty-nine occasions that they were asked the question, a persistence which is surely indicative of true lack of mechanism knowledge. The other ninety-eight pupils also said 'Don't know' from time to time, but on at least one occasion they ventured an opinion. When they did this, it was invariably with reference to a causal mechanism, suggesting that the demand characteristics of the question had worked once more as expected.

For three pupils in the Howe *et al.* sample (a 10-year-old, an 11-year-old and a 12-year-old), the causal mechanism was exclusively object-based. That is to say, objects have properties like weight which pull them down and properties like air which push them up. Whether they will float or sink is a question of which properties are in ascendance. Thus, as the 12-year-old put it 'Stones sink because they're rounded all over. So when the bottom is pushing up, the top bit is pulling down because of the weight.' By contrast, all the other pupils cited mechanisms that referred to the object–liquid interplay and in doing so followed four lines of reasoning. The first, which was used by four pupils only (two 8-year-olds and two 10-year-olds), emphasised the liquid's transformational properties, for instance 'The things in whisky make objects heavy and so they sink' and 'The water becomes part of the objects to make them float.' The remaining three approaches were more frequent and, as Table 7.6 shows, were interestingly distributed as a function of age.[3]

The first approach was a complete reversal of the scientific account in that liquids were seen as exerting a downwards force which objects may or may not be capable of countering. For most proponents, it was simply a question of the water (or whatever) pushing, pulling or sucking down. However, for some, it was a seemingly separate component like waves, bubbles, gases and germs. Typical here were the 10-year-old who said 'A big wave could bring them down like a big hand pulling them to the bottom' and the 12-year-old who claimed (on five occasions) that 'The gravity under

Table 7.6   Mechanisms of floating in liquids as a function of age
(Howe *et al.*, 1990)

	8–9-year-olds		9–10-year-olds		10–11-year-olds		11–12-year-olds	
	No. of children	Mentions per child	No. of children	Mentions per child	No. of children	Mentions per child	No. of children	Mentions per child
Liquids exert a down-wards force and objects an upwards	9	0.59	13	0.69	10	0.70	13	0.81
Liquids provide a barrier to the down-wards force of objects	12	1.63	15	1.34	12	1.33	18	1.16
Liquids exert an upwards force and objects a down-wards	9	0.41	9	0.44	8	0.57	19	1.13

the water pulls it down.' As Table 7.6 shows, forty-five pupils followed the first approach on at least one occasion, with no age differences between them ($\chi^2 = 0.69$, df = 3, n.s.). There were also no age differences in how frequently the approach was followed across the items by the forty-five users (F = 0.13, df = 3,117, n.s.).

The second approach was arguably closer to scientific wisdom in that objects were unambiguously seen as exerting a downwards force. However, rather than countering this with an upthrust, liquids were viewed as providing barriers. This was usually expressed in terms of the liquids holding or failing to hold the objects, but once more there was a minority tendency to refer to parts of the liquids rather than the whole. For two 8-year-olds and a 10-year-old, the crucial part was the liquids' 'skin'. For nine pupils all from the two oldest age groups, it was the liquids' contents, for instance 'Salt acts like a platform to keep things up.' As Table 7.6 shows, the 'barrier' mechanism was used the most frequently. However, its usage was not associated with age, neither in terms of number of pupils mentioning it ($\chi^2 = 1.76$, df = 3, n.s.) nor of number of mentions per pupil (F = 0.20, df = 3,117, n.s.).

In contrast to the other two approaches, the third approach in Table 7.6

did vary in usage as a function of age. This approach was undoubtedly the best in scientific terms in that liquids were regarded as providing an upwards force and objects a downwards. This said, none of the pupils acknowledged an upwards force from the displaced liquid, and thus the superiority of the approach was strictly limited. What the pupils called on instead was the liquid as a whole or its pressure or its components. Reflecting what we have already observed with the first and second approaches, the latter included waves, bubbles, salt and vitamins. Thus, we had 'Water in the sea has big waves which will push the bottle up', 'The salt gathers the iceberg up under-neath and pushes it up' and 'Less things will sink in milk than water because the vitamins lift things', from respectively an 8-, 9- and 10-year-old. 8-, 9- and 10-year-olds did then follow the third approach. However, where it came into its own was with the 11- to 12-year age group. As Table 7.6 shows, nineteen pupils from the oldest age group followed the approach as opposed to eight or nine from the others, a difference which with more or less equal numbers in each age group was statistically significant ($\chi^2 = 9.46$, df = 3, p<0.05). The 11- to 12-year-olds also achieved over double the number of mentions per pupil, a difference that was again statistically signif-icant (F = 2.66, df = 3,117, p = 0.05).

The age differences with the Howe *et al.* (1990) data were sufficiently striking to encourage a quest after convergent evidence from other sources. Thus, I looked also at the Howe (1991) study where it will be remembered the focus was the properties of liquids relevant to the floating of an airbed and the sinking of a plant basket. As with the Howe *et al.* study, mention of variables triggered questions to the effect of 'Why is (variable) important for floating/sinking?' However, this time, predictions/descriptions of outcome were also followed with mechanism questions. In both contexts, the pupils frequently said 'Don't know', but on the occasions that they went beyond this their responses always referred to mechanisms. For fifty-five of the 126 pupils interviewed, these mechanisms involved an upwards force from the liquids to counter a downwards force from the objects. Consistent with the Howe *et al.* data, usage of the mechanism increased with age from 15 per cent of the 6- to 7-year-olds, through 32 per cent of the 8- to 9-year-olds, 58 per cent of the 10- to 11-year-olds, 50 per cent of the 12- to 13-year-olds to 63 per cent of the 14- to 15-year-olds. This increase was statistically signif-icant ($\chi^2 = 16.39$, df = 4, p<0.01), with the two youngest groups being differentiated from the three oldest.

Although the age increase was consistent with Howe *et al.*, it took place approximately one year earlier, for the cut-off appears at 10 to 11 whereas earlier it was 11 to 12. However, this discrepancy must be seen as significant in the present context, for it reflects exactly what happened with relevant variables. Thus, it follows that in both studies the age at which upthrust became characteristic in pupils' thinking was the age at which there was a substantial increase in the use of relevant variables. Could there be a connec-tion? If so, is it one that suggests a driving role for knowledge of mechanisms? Whatever the case, what is the significance of the more primitive causal mechanisms whose usage does not change with age? Are they associated with variable generation, albeit not with relevancy? These questions will form the framework of the next part of the section.

Theoretical structure and outlying variables

In looking at the relation between causal mechanisms and variable selection, we are restricted to my own data since nothing of relevance has been produced elsewhere. Focusing then on these data, I decided to consider only the relations between the numbers of pupils using particular mechanisms and the number of distinct variables that those pupils used. Although earlier analyses have looked also at mechanism mentions and variable mentions, the results were in all cases equivalent to those based on distinct mechanisms and distinct variables. This suggested that analyses based on mentions would prove redundant.

I began with a t-test on the Howe *et al.* (1990) data to explore the hypothesis that invoking upthrust as a causal mechanism is associated with a relatively large number of relevant variables. The hypothesis was supported: the pupils who used the last of the mechanisms outlined above averaged 2.36 relevant variables against the 1.80 averaged by the other pupils, a difference that was statistically significant ($t = 2.79$, $df = 119$, $p<0.01$). Unsurprisingly given analyses presented earlier, the difference was primarily due to divergence over relevant properties of liquids, a finding which motivated a parallel analysis with the Howe (1991) data. There, it will be remembered, the sole concern was liquid variables, and there was an age-related increase in relevancies with the airbed scene. A t-test on the data from that scene revealed that the pupils who recognised upthrust were significantly more likely than the other pupils to generate relevant variables ($t = 2.41$, $df = 124$, $p<0.05$).

Thus, there is clearly some association between awareness of upthrust and selection of relevant variables, but is it an association whereby the latter is dependent on the former? As always, there are limits on what can be said without longitudinal data. Nevertheless, something can be established by following the practice of parts II and III and looking at the proportion of pupils without upthrust who pass some threshold of relevant usage compared with the proportion with upthrust who fail to pass the threshold. To the extent that the former is smaller than the latter, expansion of relevant variables is more likely to wait on upthrust than for the reverse to be the case.

With the Howe *et al.* data, the means cited in the previous paragraph (not to mention the small standard deviations) suggest 'more than two relevant variables' as the appropriate threshold for making the comparisons. At this level, 13 per cent of the pupils without upthrust passed the threshold as opposed to 56 per cent of the pupils with upthrust who failed to pass. When analysed via the Critical Ratio Test that was introduced in part II, this difference proved to be statistically significant ($z = +5.17$, $p<0.001$). With the Howe (1991) data, the threshold needs to be lower since we are talking about one test item as opposed to twenty-nine and liquid variables as opposed to an object/liquid mixture. Use vs. non-use, that is 'at least one relevant variable', would seem appropriate. Accepting this, 13 per cent of the pupils without upthrust passed the threshold as opposed to 68 per cent of the pupils with upthrust who failed to pass. Again this was statistically significant on the Critical Ratio Test ($z = +6.11$, $p<0.001$).

So far so good: we have modest evidence that knowledge of upthrust was not only associated with the use of relevancies but also imposing constraints. However, what about irrelevancies? We can probably say something here without further analysis, and it is not very encouraging. We have in this section identified a handful of mechanisms, most used by a relatively large number of pupils. By contrast, we have in the previous section identified a large number of irrelevant variables, most used by a handful of pupils and many by just one. It is virtually impossible to think how we could have made both observations if all irrelevant variables were generated by mechanisms. The best that we could hope for is that some irrelevant variables are derived from mechanisms; the chances of their all being so derived seem virtually zero.

Taking this for granted, the best chance for dependency on mechanisms probably lies with the small number of irrelevant variables which recur across children. In which case, the variables to focus on in the Howe *et al.* (1990) data are the nine which stood out as being used by at least 25 per cent of the sample. Expressed as facilitators of floating, these variables were the airiness, woodenness, roundness, non-metality and non-wateriness of objects and the lightness, stillness, movement and depth of liquids. I carried out chi-square analyses to see whether there was an association between use of these variables and reference to the three most frequent causal mechanisms. Most of the analyses yielded non-significant values, but there were two exceptions. The pupils who believed that liquids exert a downwards force were significantly more likely than the other pupils to see object airiness as a facilitator of floating ($\chi^2 = 5.94$, df $= 1$, $p=0.01$). Moreover, 49 per cent of the pupils who subscribed to 'downthrust' did not call upon airiness as opposed to 29 per cent who called upon airiness but did not subscribe to downthrust. On the Critical Ratio Test, this was a significant indicator of variable dependency on mechanisms ($z = +2.27$, $p<0.05$). In addition, the pupils who saw liquids as setting up barriers were significantly more likely than the other pupils to see liquid movement as facilitative of floating ($\chi^2 = 9.77$, df $= 1$, $p<0.01$). Again, subscription to the mechanism was more likely in the absence of the variable than was the reverse, the relevant percentages being 47 per cent and 25 per cent respectively. The difference proved once more to be statistically significant when analysed on the Critical Ratio Test ($z = + 2.59$, $p=0.01$).

Comparable analyses were not carried out on the 1991 data, partly because the data related to liquid variables only and partly because the seemingly crucial liquid variable, movement, was not used with sufficient frequency. Concentrating as a result on the above, we do, I think, have some evidence that variable selection was influenced by causal mechanisms. All three of the major mechanisms were associated with the selective use of certain variables, downthrust with object airiness, barriers with liquid movement and upthrust with relevant properties of liquids. Moreover, the nature of the association was, without exception, what would be expected if the mechanisms were making the running. This is particularly interesting when we remember that even though upthrust was relatively rare until after 10 or 11, downthrust and passive resistance were common at all ages studied. After all, the implication from this would seem to be that generativity is available from a very early age, bringing object flotation into line with the transfer of heat. Thus it would not simply be that object flotation resembles heat transfer over the age profiles for variable and mechanism understanding, a point signalled earlier. It would also be that there is a resemblance over the mechanism–variable relation. All this would be somewhat ironic, for it would separate object flotation from propelled motion when from the received perspective they are closely related.

Having said all this, I am not claiming that the data 'prove' that object airiness, liquid movement and the relevant properties of liquids were dependent on their associated mechanisms. I am simply stating that currently this is the most plausible gloss and that this suggests parallels on one level with heat transfer. However, even if all this is accepted, we must not lose sight of the fact that object airiness, liquid movement and the relevant properties of objects are a tiny subset of the total variables used, and contingency on mechanisms seems highly unlikely in the remaining cases. Reference has already been made to the large number of variables that were unique or nearly unique to single pupils, in comparison with the small number of mechanisms that were in most cases used repeatedly. There is no suggestion that contingency implies exact correspondence of variables across pupils who share mechanisms, nor is there any suggestion that it denies a fulsome range of variables. However, the discovery that pupils can share mechanisms yet show virtually no correspondence over variables and a massive range besides seems impossible to reconcile with contingency. In addition though, there were the variables which, though common, were not associated with particular mechanisms. These included the seven irrelevant variables additional to object airiness and liquid movement which were used by at least 25 per cent of the Howe *et al.* sample. However, they also included object lightness, a relevant variable. It was used by every pupil in the Howe *et al.* sample apart from one 8-year-old, and hence was self-evidently not associated with a particular mechanism.

The trouble is that once we recognise these outlying variables, we do not merely have to acknowledge limits on generativity. We also have to accept that the parallels between object flotation and heat transfer are less exact than they were beginning to seem. As noted in part II, very few heat transfer variables lie beyond the scope of mechanisms. Why, then, do literally hundreds of them do this with object flotation? Our emergent model could provide an answer, with reference as it happens to the mechanisms which, from the age trends, children must begin with. We have just seen that with object flotation these mechanisms will be downthrust and/or barriers. In part II, we established that with heat transfer the initial mechanism is likely to be transfer without transmission.

The key difference between the mechanisms in the two topic areas is that downthrust and barriers are *reversals* both of the everyday upthrust which sometimes replaces them and of the received scientific account. Transfer without transmission is a *partial* version of transfer via transmission, the mature everyday mechanism, and transfer via transmission is in turn a partial version of conduction. The variables generated by reversals of reality are likely to be far more wayward than the variables generated by partial truths. As such, they are likely to receive far more negative feedback when they are used in action or referred to in dialogue, resulting, if our model is correct, in abandonment, replacement and from there prolonged cycles of unconstrained induction. To express it differently, the point from our model's perspective is that when mechanisms are as poor as downthrust and barriers, their generative status may survive but the variables which they generate will be under pressure from feedback. Once feedback starts driving its wedge, the floodgates are opened for variables which are independent of mechanisms.

If the above account is correct, children should respond differently to tasks which invite them to determine heat transfer variables in response to feedback than they do to tasks which invite them to determine object flotation variables. In the case of heat transfer, they should always refer feedback to mechanisms, for example 'There can't be more cold in a big block of ice because it's the same temperature as that tiny cube.' Hence, improved variables should be contingent on improved mechanisms. In the case of object flotation, children should sometimes refer feedback to mechanisms, for example 'The air's helping it fight against the waves', but frequently they will not be able to do this because the mechanism–variable relation has been severed, hence for example 'It can't be anything to do with holes because that one floated and it's got holes', an interpretation which is focused exclusively on variables. As a consequence, improved variables will not be contingent on improved mechanisms, but rather on adequate syntheses of the variable–feedback relation.

Evidence was presented in part II to show that children do respond to heat transfer tasks in the predicted fashion. This evidence emerged from a study by Howe *et al.* (1995a) where it will be remembered pupils worked collaboratively on specially designed tasks. When the tasks encouraged both 'rule generation' (deciding what is important for heating and cooling) and 'critical test' (exploring variables through ordered sets of items), the pupils showed clear signs of: (a) calling on mechanisms to resolve disagreements over which variables could in the light of feedback be endorsed; and (b) treating the resolution of any difficulties with mechanisms that were thereby exposed as a prerequisite for the resolution of disagreements over variables. Remembering that this is what was observed with the transfer of heat, the question is whether anything different would be observed with object flotation, and if it would, whether it would be consistent with the present proposal. Finding out is of some significance in the context of this chapter. Since the proposal is derived from our model, evidence supporting it would simultaneously constitute evidence supporting the model.

It is fortunate then that something approximating supportive evidence appears in the study by Tolmie *et al.* (1993) which was mentioned briefly in part II. This study involved 128 pupils aged 8 to 12, working in groups of four on tasks which encouraged both rule generation and critical test, one of these, or none. Progress on variables resulting from the tasks was ascertained by change from individual pre-tests prior to the tasks to individual post-tests four and eleven weeks afterwards. In detail, the pre-tests consisted of Parts I and II of the Howe *et al.* (1990) interviews. The post-tests consisted of analogous items. The group tasks focused on the flotation in a tank of water of six three-item sets. In the case of the 'plus critical test' groups, these were: (a) boxes made of wood, plastic and metal; (b) plastic objects that were round, flat and curved; (c) bottles that were hollow, filled with water and filled with a solid; (d) wooden blocks, identical in size but different in weight; (e) wooden blocks, identical in weight but different in size; and (f) a random assortment. In the case of the 'minus critical test' groups, the same items were used but this time all sets were drawn up at random. For each set, group members were invited to make private predictions as to whether all, some or none of the items would float rather than sink, share their predictions and come to an agreement, test the agreed prediction by immersion in the water, and interpret jointly what transpired. 'Plus rule generation' groups then received a further instruction to look across all the items tested so far and write down the 'thing that is important for floating and sinking' and the 'things which don't matter'.

As with the Howe *et al.* (1995a) study, the groups were videotaped while they worked on the tasks, and indices derived from the videotaped dialogues were correlated with pre- to post-test change. It turned out that with the eight groups who experienced rule generation plus critical test, disagreements over variables did give rise to references to mechanisms ($r = +0.85$, $df = 6$, $p<0.01$). However, this was of no consequence for pre- to post-test change. The sole predictor of pre- to post-test change was the adequacy of the rules produced in response to the rule generation instructions ($r = +0.68$ between rules scored 0 = Level I to 4 = Level V and pre- to post-test change in percentage of relevant variables, $df = 6$, $p<0.1$). The correlations between 'rule scores' and both disagreements over variables and references to mechanisms fell a long way short of statistical significance. However, a reasonably good predictor of rule scores (and by far the best predictor amongst the dialogue variables) was 'chairing', ($r = +0.65$, $df = 6$, $p<0.1$) a form of dialogue which involved tying things together to move on with the task, for example 'Well, we all think being light is important, so let's write it down', 'We'd better vote on shape', and 'Remember what Andrew said, those two are the same size but that one's lighter so that's why it floated. It's being light for the size that counts.'

The picture is, then, very different from that observed for heat transfer by Howe *et al.* (1995a). This is despite the procedural similarities between the two studies. However, the differences are not unintelligible, for while the heat transfer results mesh with what was hypothesised there, the object flotation results seem to square with our current proposal. The references to mechanisms in the context of disagreement suggest that mechanisms were seen as relevant to the derivation of variables, and in truth we should not be surprised. We do not only have the contingency analyses of a few paragraphs earlier. We also have the fact, implicit for many pages, that the pupils in the Howe *et al.* (1990) and Howe (1991) studies were prepared to respond to 'Why is (variable) important?' by providing a mechanism. This, as we saw in part III was not the case for propelled motion.[4] In addition, pre- to posttest change in the Tolmie *et al.* study was predicted directly by rule scores and indirectly by chairing. The latter can be interpreted as attempts to synthesise variables and the former as reflections of how well syntheses were done. Thus, both seem indicative of what was hypothesised to be a consequence for the learning process of variable–mechanism severance.

Structure and content

The picture that is emerging is of a structural similarity between object flotation and heat transfer which is torn apart by content. Both are, at heart, theoretically structured topic areas, and become so from a very early age. However, the content of the heat transfer mechanisms is, even at the beginning, sufficiently orthodox to preserve the close mechanism–variable alignment which theorising implies. The content of the earliest object flotation mechanisms is so far removed from orthodoxy that the generated variables are soon under pressure and, as a consequence, replaced and severed from mechanisms. With both heat transfer and object flotation, there will be little progress over the content of mechanisms for a number of years. However with heat transfer, the situation as regards variables will be relatively stable. With object flotation, there will be frenetic shifts of opinion, and probably repeated cycles of rejection and acceptance of specific variables.

Thus, on the surface, there will be marked differences between the two topic areas, but the point being made here is that underneath there is a fundamental unity. Moreover, both the surface differences and the underlying unity are interpretable on our model. The unity between heat transfer and object flotation is not however shared with propelled motion, for the central claim of part III was that the conceptualisation of speed prevents theoretical structure from becoming established. As noted already, this separation of flotation from motion is highly paradoxical when from the Newtonian perspective flotation is the achievement in a fluid of a terminal velocity (and hence speed) of zero. It also raises distinct questions about what will happen when the fluid is no longer a liquid but is rather a gas. Will conceptual structure follow what we have established in this chapter for flotation in liquids or will it follow what we observed in the previous chapter for vertical fall? In the next chapter, we shall see what the literature has to say.

8 Flotation in gases or failure to fall

Evidence presented in the previous chapter suggests that flotation in liquids is associated with a partially theoretical structure. Variables are generated by mechanisms to be sure, but at the same time many variables lie beyond the scope of mechanisms. Seeking to interpret this, the previous chapter called upon the lack of orthodoxy displayed by many children in their choice of mechanisms. This, it was proposed, should lead to waywardness over variables, negative feedback on usage, unprincipled substitution of new variables, and gradual severance from mechanisms. The scenario was contrasted with the situation relating to the transfer of heat, where the more 'respectable' mechanisms should (and seemingly do) result in a tighter theoretical structure. Yet while the contrasts were spelled out, the previous chapter recognised the core similarity between flotation in liquids and heat transfer which stemmed from their mutual reliance upon theory. This, it was argued, not only brings the two topic areas together but also sets them apart from propelled motion, as discussed in part III. Since flotation in liquids is, from the received perspective, a specific instance of propelled motion, this is not merely ironic; it suggests also that the segmentations which are fundamental within everyday physics are very different from those which apply within orthodox science.

Having made such points, the previous chapter raised the question of what they imply for flotation in gases. From the received point of view, an object which floats in a gas is an object which fails to fall. If this is also the case within everyday physics, it suggests that flotation in gases will be aligned with unsupported projected motion as analysed in the latter half of part III. However is this the case? Perhaps flotation in gases is aligned with flotation in liquids, and hence separated from projected motion. The aim of the present chapter is to shed some light, and by way of introduction, it should be noted that there is a major difference between gases and liquids which could have consequences for how flotation is viewed. This is that the density of gases varies with pressure while, to all intents and purposes, the density of liquids does not. Since density is the key factor in object flotation, the implication is that the propensity to float in gases will change as pressure changes while the propensity to float in liquids will not do this.

The potential significance of the difference between gases and liquids becomes clearer once we appreciate that since atmospheric pressure decreases as we rise above the earth's surface, flotation in air will not be all or none. A feather released from a high-flying aircraft may fall some way until the density of the air is sufficient to create a compensatory upthrust. A balloon released from the ground may soar a few hundred feet until the density of the air is insufficient to create the needed upthrust. This is in marked contrast to, say, water. Even though water pressure increases as we move below the surface, a harpoon released from a submarine is no more (or less) likely to float than one released from a boat. Since air is the major gas of our experience as water is the major liquid, the consequence of the contrast is to make flotation in gases phenomenologically more dynamic than flotation in liquids. Indeed, the dynamism is such that outside the science laboratory we either avoid the terms 'floating' and 'sinking' when talking about air or use them in a reversed sense, as with 'the leaves float gently downwards'.

All of this must make it seem highly debatable whether our everyday physics of flotation in gases will be closely related to our everyday physics of flotation in liquids. Indeed, the dynamism may on the face of it appear to increase the probability that flotation in gases will turn out to be more closely aligned with vertical fall. In any event, there are good reasons for reviewing the relevant literature, and this is what the present chapter will do, looking as always at variables, mechanisms and their interrelation before making some broader theoretical points.

PRESSURE, DENSITY AND VARIABLE SELECTION

Although the focus must (and will) be on variables, mechanisms and their interrelation, there is a prior question which we need to address: do children appreciate the relationship between pressure and density as it pertains to liquids and gases? After all, if they do appreciate the relationship, they will have the tools to understand and therefore transcend the phenomenological differences. If they do not appreciate the relationship, a linkage between liquids and gases could hardly be predicted, and would in fact require explanation. Thus, to provide a framework for interpreting what is to follow, the present section will start with data from my 1991 study which bear on the pressure–density relationship. Having reported the data and applied them to the phenomenology of flotation, the section will turn to the chapter's first major theme, the variables relevant to flotation in gases.

Pressure and density

I included two questions in the Howe (1991) study to see whether the pressure–density relation was appreciated by children. The first question was presented in introduction to the airbed scene described in the previous chapter. Once it had been established that the 126 pupils interviewed saw the airbed as about to sink, they were asked whether and how the water would change as the airbed (and a swimmer perched on it) went deeper and deeper. The second question related to a photograph where the strings of three helium balloons were wedged under a brick which prevented them from moving. The pupils were invited to imagine the brick being kicked away and the balloons soaring upwards. They were then asked whether and how the sky would change as the balloons went higher and higher.

The pupils responded to the questions in one of five ways: (a) no changes were identified, that is the pupils either denied that the water/sky would change or did not know what the changes would be; (b) person-oriented changes were identified, for example 'It'll be more spookier' and 'Your nose'll get stuffed up'; (c) physical changes apart from pressure and density were identified, for example 'It'll get darker, colder and windier' and 'The sky will get bluer'; (d) density and/or pressure were referred to (not necessarily using the terms) but not properly grasped, for example 'Air pressure will increase' and 'Under the sea, it's darker and heavier';[1] and (e) changes in pressure and (for sky) density with altitude were understood, for example 'The air pressure and density will get less.'

Many pupils used more than one response type. However because the types lie on a clear continuum in terms of adequacy and because these is no need here to represent the breadth of ideas,[2] the data could be simplified by treating each pupil as having given the most adequate response only. It is on this basis that the pupils are classified in Table 8.1, where it is clear that at no age level do the majority of pupils have a firm grasp of what is at stake. There are however signs of improvement with age, not altogether surprising given that the two oldest groups had studied the topic at school. To test the statistical significance of the improvement, the pupils were assigned scores from zero to four depending on the adequacy of what they had said and the age groups were compared by analysis of variance. With both the airbed scene and the balloons, the age effects were highly significant (F for airbed = 9.39, df = 4,121, $p<0.001$; F for balloons = 18.79, df = 4,121, $p<0.001$). With the airbed scene the three youngest groups obtained significantly lower scores than the two oldest. With the balloons, the 6- to 7-year-olds and the 14- to 15-year-olds differed significantly from the middle three age groups and from each other, the former being the least advanced and the latter the most. There was also a significant correlation between the scores for the two scenes ($r=+0.46$, df=124, $p<0.001$), attesting to a degree of association.

It is difficult to find anything in the literature to compare these results with. Clough and Driver (1985b) asked eighty-four 12- to 16-year-olds what would happen to the pressure on first a goldfish and second a submarine as the depth of water increased. Consistent with Table 8.1, they found an improvement with age in the percentage giving the correct answer. However, the improvement was from 60 per cent at 12 to 88 per cent at 16, a higher absolute level of performance than that depicted in Table 8.1. The difference

Table 8.1 Appreciation of the effects of altitude and depth on pressure and density (Howe, 1991)

	6–7 year-olds		8–9 year-olds		10–11 year-olds		12–13 year-olds		14–15 year-olds	
	Airbed	Balloons	Airbed	Balloons	Airbed	Balloons	Airbed	Balloons	Airbed	Balloons
No change	11	9	7	3	8	1	7	2	2	0
Person-oriented change	7	9	7	2	4	3	1	3	0	0
Physical change (excl. pressure and density)	8	6	6	12	4	12	7	6	6	3
Inaccurate pressure and density change	0	2	3	8	4	7	2	12	8	17
Accurate pressure and density change	0	0	2	0	4	1	7	1	11	7

probably reflects one or both of two considerations: (a) pressure was signalled explicitly in Clough and Driver's interviews while mine gave little direction; and (b) pressure knowledge was scored separately by Clough and Driver while I combined it with density. For present purposes, it is the combination that is relevant. As stressed already, the phenomenological differences between flotation in liquids and flotation in gases result from differences in the pressure–density relation. Thus if children are to transcend phenomenology and appreciate the underlying links, secure knowledge of both pressure and density would appear to be needed. On the evidence of Table 8.1, such knowledge is the exception not the rule in the 6 to 15 age group.

Confirmation that children seldom transcend phenomenology comes from the work of Rodrigues (1980). As explained earlier, Rodrigues interviewed fifteen pupils aged 7 to 13 about cartoon instances of object flotation. In addition to the instances relating to liquids there were instances relating to gases, and at one point in the interviews, Rodrigues requested explicit comparisons. She indicated two cartoons similar to the ones depicted in Figure 8.1, namely an air-filled ball springing out of water and a helium-filled balloon soaring into the sky and asked whether the cartoons had anything in common. Seven pupils said 'No', arguing for example 'It's not the same thing because this one has gas and this one has air' (Rodrigues, 1980: 67). Four said 'Yes' but gave spurious reasons, for example 'It's sort of the same, but this ball goes up to the surface of the water, but it won't probably go up in the air' (Rodrigues, 1980: 67). Only four pupils said things like 'Sort of, because this is lighter than air and this lighter than water' (Rodrigues, 1980: 67) which provided convincing evidence that the link was grasped.

Variables and the centrality of weight

It is interesting that in the example just given the pupil was using relatively sophisticated variables to justify the judgment, variables that would be coded at Level IV in the scheme used earlier. Such variables were rare with liquids, and we must wonder about their frequency with gases. Rodrigues does not provide the requisite information, but she alludes to variable selection at several points in her paper and something can be gleaned about what was going on. It is obvious that 'lightness for the air' must have been rare. However, it is equally obvious that lightness in an absolute sense was all pervasive. For most of Rodrigues' sample, objects float in air when they are light and sink to the ground when they are heavy. With liquids, lightness was also important as a facilitator of floating. Indeed, it was probably the most frequently mentioned variable in every one of the studies outlined in the previous chapter. However, it was not unique, and it was indeed just one of over 200 distinct variables in the Howe *et al.* (1990) data. Are the equivalents of these other variables missing when we talk about gases, or is there some-

Figure 8.1 Cartoons equivalent to those used by Rodrigues (1980)

thing about Rodrigues' methodology and/or reporting that is eclipsing the variability?

For the former and against the latter is the fact that attempts to tap heterogeneity directly by pinpointing other variables have often found themselves returning to weight. For example, Brook and Driver (1989) gave 100 5- to 16-year-olds the opportunity to discuss whether having air inside was a relevant issue. Some pupils thought that it was, arguing in the words of a 16-year-old that 'Balloons with air float' (Brook and Driver, 1989: 31). However, many pupils (particularly at the older age levels) viewed air as having 'negative weight', with inflated balloons believed to weigh less than flat ones. Thus, when pupils suggested that air helps floating, it was because air was seen as reducing weight.

A similar message comes from the studies by Noce *et al.* (1988) and Ruggiero *et al.* (1985) discussed in part III. It will be remembered that in the course of these studies, over 400 individuals whose ages ranged from early teenage to adulthood were questioned about: (a) what would happen to a spanner if an astronaut let go of it while standing on the moon; and (b) what would happen to a stone resting in a glass container if the air was evacuated. The point of relevance here is that ambient atmosphere was identified as a variable for subjects to consider and there were differences in whether floating or sinking was anticipated. However, without exception, responses were dictated by beliefs about the impact of 'airlessness' on weight. If the participants thought it had no bearing, they predicted sinking. If they thought (as many did) that it reduced weight, they predicted floating. In due course, we shall discuss why weight and airlessness were sometimes interrelated, an issue which will turn out to be of great relevance to our central concerns. For now, the only point to note is the simple one: even when the focus was on the wider context, object weight still proved to be crucial.

Clearly then, weight is important but is it really the be all and end all? In contrast to the work discussed in the previous chapter relating to liquids, published research into flotation in gases does not appear to confront children with the really awkward cases, that is heavy objects which float. It would be interesting to find out what children said about hot air balloons, gliders and (if they think it floats as some probably do) the sun. It would also be interesting to explore the full range of atmospheric conditions, for instance turbulence, pollution and moisture, rather than just airlessness. I had the latter possibility in mind when designing the Howe (1991) study for I included two photographs which provided contexts for quizzing the pupils about the atmosphere. The first was the balloons scene which was described earlier; the second was a scene from a children's cartoon where two gnomes were flying on a magic carpet.

With the balloons scene, the pupils were invited to imagine a competition to see which balloon would go highest. They were asked in a general way whether anything could be done to help one balloon win. Then, depending on what they had already said, they were asked whether a higher starting point (up a tree or in an aeroplane), a change of weather or a higher level of pollution would make a difference. With the magic carpet scene, the younger pupils were invited to imagine that the gnomes wanted to land. (The older pupils were invited to focus on the landing of 'something else that flies without an engine' and usually selected a glider.) Paralleling the balloons scene, there was a general question about whether anything could be done to help, followed by specific questions about the altitude of the land below, the weather and the level of pollution.

Far from denying that properties of the sky were relevant and/or locating all the variance in object weight, the pupils produced twenty-nine different atmospheric variables with the balloons scene and twenty-one with the magic carpet. As it turned out, these variables were clearly divisible into the 'idiosyncratic' and the 'consensual': with the balloons scene, twenty-four variables were used by five pupils or fewer, and five variables were each used by seventeen pupils or more (up to seventy-three pupils in one case); and with the magic carpet scene, seventeen variables were used by ten pupils or fewer and four variables were each used by seventeen pupils or more. It was immediately apparent that usage of many of the consensual variables changed with age. With the balloons scene, inverted U-distributions peaking at 10 to 11 years were observed with clean air ($\chi^2 = 13.03$, df = 4, p<0.05) and no rain ($\chi^2 = 15.58$, df = 4, p<0.01) as helps to flying high. Wind decreased with age as a help ($\chi^2 = 10.42$, df = 4, p<0.05) and no wind increased ($\chi^2 = 16.34$, df = 4, p<0.05). With the magic carpet scene, both no wind ($\chi^2 = 10.56$, df = 4, p<0.05) and dirty air ($\chi^2 = 13.28$, df = 4, p=0.01) showed inverted U-distributions as aids to coming down. Across the scenes then, there was exact equivalence over the age distribution for 'Clean air implies floating and dirty air implies sinking' and near equivalence for 'Wind implies floating and no wind implies sinking.'

None of the consensual variables could be called 'relevant' in the sense of the previous section. Pollution used in a reversed way, that is 'Clean air implies sinking and dirty air implies floating', would have been acceptable if it also carried the sense of denseness or thickness. However, 'reversed' pollution (let alone the association with denseness or thickness) was most definitely non-consensual. It was occasionally used by the 14- to 15-year-olds, as for example when one claimed that 'The dirt particles will make the air dense so the balloons will go higher', but such use was rare. It was certainly not sufficient to make for significant age differences over relevant variables: with neither the balloons scene nor the magic carpet did the usage of relevancies change with age. It remained at a consistently low level.

THEORIES IN THE OBJECT–GAS INTERPLAY

When we considered liquids in the previous chapter, we did, in contrast to the above, find an age-related improvement in the use of relevant variables. Moreover, it was an improvement that was continued when, as with the above, we focused on fluid properties with features of objects more or less excluded. Since the age-related improvement in relevant properties of liquids was associated with an enhanced understanding of causal mechanisms, the failure to find a parallel improvement with the relevant properties of 'sky' may signal differences in how knowledge is structured. Are the differences to be found? The present section will explore the question in detail, looking first at the causal mechanisms which apply with gases, then at their implication in variable selection, and finally at the broader theoretical issues which are thereby raised.

Downthrust and upthrust in gases

The first step is then to consider what mechanisms children subscribe to in relation to gases. The work of Bliss *et al.* (1989) and Hayes (1979) implies that discrepancies with liquids will be found here too. In both pieces of work, the belief that falling through air is 'natural' in the absence of support is claimed to be a central tenet of everyday physics. The implication is that falling through air does not require explanation beyond reference to support being absent. By way of endorsement, the twenty-six pupils that Bliss *et al.* interviewed referred to lack of support and very little else when interpreting the falling displayed in comic cartoons. The crucial point to be drawn here is that if Bliss *et al.* and Hayes are correct and falling is genuinely treated as natural in the absence of support, then children should not need extraneous 'pushing down' forces to provide an explanation. Thus, the downthrust mechanism that played a central role in relation to liquids should seldom find an equivalent in the context of gases. Equally, if falling is treated as natural, then failure to fall should be the occasion for considerable surprise and cry out for explanation. Hence, the upthrust mechanism that with liquids only became prevalent around 10, 11 and 12 may prove more salient when thinking about gases.

Intrigued by the hypotheses that seemed to be implied, I arranged for the Howe (1991) research to include questions with the balloons and magic carpet scenes that were directed at mechanisms. As with the other scenes, these were questions which followed up mentions of outcomes with 'How is (outcome) achieved?' and questions which followed up mentions of variables with 'Why is (variable) important?' As it happened, thirty-five of the pupils referred to the pulling down properties of air or its components by way of an answer. Six of these pupils were in the 6 to 7 age group, ten in the 8 to 9, eight in the 10 to 11, five in the 12 to 13 and six in the 14 to 15. Although this suggests some concentration in the middle years, the age difference was not statistically significant (χ^2 = 4.04, df = 4, ns). Indeed at all ages, the pupils referred to air, air pressure, dirt, smog, pollution, thunder, rain and chemicals as the pulling down properties. The only thing that differentiated the older pupils from the younger was the occasional spurious science amongst the former. For instance, one 14-year-old claimed 'The CO_2 puts things down' and another 'Gravity and air pressure will pull it down.'

The point is that with thirty-five pupils referring to the pulling down properties of air, a sense of downthrust was nearly as prevalent with flotation in air as it was with flotation in water. This seems to go against what Bliss *et al.* and Hayes would predict. How about upthrust? It is possible to hypothesise from Bliss *et al.* and Hayes that this will be more prevalent for gases than it was for liquids. Yet my 1991 data give no grounds for thinking that this is the case. It will be remembered that fifty-five of the pupils viewed the water in the airbed and plants scenes as exerting an upwards force. Only thirty-seven held an equivalent view about the sky in the balloons and magic carpet scenes. These thirty-seven were concentrated at the younger age levels, with 35 per cent of the 6- to 7-year-olds, 44 per cent of the 8- to 9-year-olds and 33 per cent of the 10- to 11-year-olds mentioning an upwards push as opposed to 17 per cent of the 12- to 13-year-olds and 19 per cent of the 14- to 15-year-olds. Although the age difference was not statistically significant (χ^2 = 6.15, df = 4, ns), it is none the less curious for it is an exact reversal of what was observed in the previous chapter for liquids. There, mention of upthrust increased substantially around 10 or 11 and stayed high thereafter.

Seeking an explanation, it is tempting to bring two literatures to bear. The first stems from Piaget (1930), in particular the research reported in that book concerned with children's conceptions of the 'nature of air'. This research involved a series of informal demonstrations, for example palms clapped, a ball deflated and an object swung round, all creating 'perceptible' air. Children were asked where the air was coming from. At the younger ages, the air was seen as emanating from the focal object, the palm, the ball or whatever, sometimes in collaboration with a vague 'outside'. It was not until much later that the children recognised the air as being displaced from the atmosphere. This suggested to Piaget that for young children air is not omnipresent but dependent on dynamic events. Not surprisingly given what we have already learned about the 1930 book, the dynamism was related to human action and from there to personal volition.

Writers subsequent to Piaget have explored (and largely confirmed) the denial of omnipresence in favour of dynamism without being overly inter-ested in the putative volition. For example, Bar (1986) asked forty Israeli children aged 7 to 9 what was the source of the air experienced when paper was waved close to their faces. About half of the sample said that the air was created by the hand's movement, rather than pre-existing within the room. Borghi *et al.* (1988) interviewed thirty-five Italian children aged 6 to 8 about the presence of air in the classroom, a container and a (swollen) plastic bag. Some 40 per cent of the 6- to 7-year olds and 25 per cent of the 7- to 8-year-olds believed the presence of air to be dependent on movement. Brook and Driver (1989) found that even the youngest children in their study of 100 British pupils aged 5 to 16 agreed that 'air is everywhere'. Nevertheless, when questioned more closely about a jar with lid, it became apparent 'that early ideas about air are primarily associated with its motion, hence air may be seen to exist in open spaces where there are draughts and breezes but not in closed containers' (Brook and Driver, 1989: 23). The point is reinforced by Séré (1986) when she reports that even at 11 years, the majority of a sample of French pupils believed that you have to hold an open bottle in a draught to fill it with air.

There seems little doubt then that air is a dynamic quantity for many young children. However, if this is the case, it is not surprising that so many of my sample talked about air pushing up. Indeed, the recognition that air can push up should not preclude the possibility of it also pushing down, for the direction of movement is presumably seen as fluid. One 8-year-old acknowledged this explicitly in the 1991 study, arguing that 'Stormy weather will sometimes help the balloons go up, but sometimes it won't. It depends on which way the wind's blowing.' In this context, the upthrust mentioned by over one-third of the sample in the 6 to 11 age group has to be recognised as extremely primitive. It is far removed from the upwards force attributed to liquids from 10 to 11 onwards and no wonder that it appears to diminish at around that age. As noted already, the 12- to 15-year-olds were less than half as likely as the younger pupils to mention pushing up.

Although the difference between the younger and the older pupils can be understood as a partial (at best statistically non-significant) decline in primitive upthrust, it nevertheless raises a number of questions. In particular, it signifies a reduced commitment to upthrust in any form, not just the primitive one. Thus, we must ask why the older pupils did not switch on any substantial scale to a more advanced version of upthrust, akin perhaps to what many of them were using for liquids. This is where the second literature alluded to earlier comes in, for it indicates that once air is seen as omnipresent, it is also seen as passive. This literature includes further aspects of the Brook and Driver (1989) and Séré (1986) research, together with the different but related work of Clough and Driver (1985b, 1986) and Séré (1982). In all cases, pupils were presented with familiar (or fairly familiar) instances of suction. These included drinking water through a straw, unblocking a sink with a plunger, pushing a suction pad onto a wall and draining liquid through a syringe. Explanations were solicited as to how the effects were achieved. A range of answers were given, including reiteration of the human action and/or the 'need' to fill a vacuum. However, the main point for present purposes is that virtually no pupils referred to atmospheric pressure and/or forces related to this. As Séré (1986) puts it with more than a hint of decisiveness 'For the pupils, atmospheric air exerts no forces' (Séré, 1986: 420).

Downthrust, upthrust, pollution and weather

The two mechanisms sketched in the preceding paragraphs more or less exhaust the pupils' ideas. There was in particular little evidence of 'air as a barrier', equivalent to the most popular of the mechanisms identified in the previous chapter for liquids. In fact, only eleven pupils talked in these terms, saying things like 'The exhaust fumes would block it' and 'The black clouds will stop you from coming down'. Thus, as regards mechanisms, we are confronted with two equally primitive notions, and the question is what role do these play in variable selection? Paradoxically, the primitiveness of the mechanisms gives us grounds for thinking that their role may not be out of step with what we observed for liquids. With liquids, the primitive mechanisms were associated with variables which related to the irrelevant properties of fluids; variables relating to relevancies waited on something that was more advanced. Thus, it is suggestive that my 1991 data indicated that variables relating to relevant properties of fluids are rare in the context of air. It is even more interesting that neither of the primitive mechanisms was cited by the nine pupils who showed good understanding at the level of variables. These were the pupils who mentioned 'weight relative to the air' with at least one of the scenes. When questioned as to why this was important, they studiously refused to be drawn on mechanisms. Instead they either claimed not to know or responded in a circular fashion often with explicit reference to density. Thus, we had 'The pollution will make the balloons go higher.' (Why?) 'Because the sky will be heavier compared with the balloons.' (Why is that important?) 'Because the sky will be denser' or 'Because the meeting place of the densities will be higher.'

So the primitive mechanisms were not associated with relevancies, but were they associated with irrelevancies? To find out, I computed the association between each of the mechanisms and the variables identified earlier as consensual. It will be remembered that there were nine such variables, five from the balloons scene and four from the magic carpet. Thus, eighteen associations were scrutinised in total, and six produced statistically significant results. These are presented in Table 8.2. As Table 8.2 shows, the picture as regards 'downthrust' was entirely consistent across the scenes. The pupils who believed that air or its components exert a downwards force were significantly more likely than other pupils to believe that clean air and no rain helps things float and dirty air and rain helps things sink. Moreover in all cases subscription to the variable without the mechanism was less common than subscription to the mechanism without the variable, suggesting that the mechanism may have been generative.

The message with regard to 'upthrust' is harder to interpret. With the balloons scene, both 'wind' and 'no wind' were significantly associated with use of the mechanism. With the magic carpet scene, neither were. This is despite the fact that 'wind' and 'no wind' were both consensual variables with that scene. Focusing then on the balloons scene, it seems from Table 8.2 that the pupils could believe in wind helping floating no matter whether they subscribed to upthrust or not. However, they were very unlikely to subscribe to upthrust unless they viewed wind as helping. This, for once, suggests a dependency of mechanisms on variables. On the other hand though, beliefs in no wind being helpful were incompatible with upthrust, and since upthrust was a relatively early acquisition declining with age and no wind a relatively late one it could be that acquisition of the variable was dependent on loss of the mechanism. This is a constraining influence of mechanism, albeit of a relatively weak kind.

My study focused on children's appreciation of the fluid variables relevant to flotation, and similar work is needed relating to objects. Nothing systematic currently exists, yet if we relate the findings of Noce et al. (1988) and Ruggiero et al. (1985) discussed in the present chapter to those discussed in part III we encounter something very interesting. The findings of concern to the present chapter are that airlessness in the environment is sometimes thought to reduce object weight and air is sometimes thought to increase it. The findings of concern to part III (and endorsed by Ameh, 1987 and Watts, 1982) are that air can produce a downwards force in collaboration with gravity, via gravity or by virtue of being gravity. Putting the two points together, the forces are creators of weight, a view made explicit by the 25 per cent of the sample who argued that gravity (defined as above) is responsible

Table 8.2   Relations between mechanisms and variables: flotation in gases (Howe, 1991)

Air exerts a downwards force

| | *Balloons scene* | | | *Magic carpet scene* | |
	Clean air helps floating			*Dirty air helps sinking*	
	- Variable	+ Variable		- Variable	+ Variable
- Mechanism	75	16	- Mechanism	83	8
+Mechanism	21	14	+Mechanism	23	12

$\chi^2 = 7.00$ p <0.01 $\chi^2 = 12.30$ p <0.001

	No rain helps floating			*Rain helps sinking*	
	- Variable	+ Variable		- Variable	+ Variable
- Mechanism	82	9	- Mechanism	84	7
- Mechanism	23	12	+ Mechanism	22	13

$\chi^2 = 10.83$ p = 0.001 $\chi^2 = 16.42$ p = 0.0001

Air exerts an upwards force

| | *Balloons scene* | |
	Wind helps floating	
	- Variable	+ Variable
- Mechanism	51	38
+ Mechanism	3	34

$\chi^2 = 25.83$ p = 0.0001

	No wind helps floating	
	- Variable	+ Variable
- Mechanism	74	15
+ Mechanism	36	1

$\chi^2 = 4.72$ p < 0.05
(df = 1 in all cases)

for weight. The crucial implication is that if air/gravity are used to explain weight, then mechanisms are generating the variables in a very real sense.

Flotation in gases or vertical fall?

Thus, as with flotation in liquids, we have a situation where mechanisms could be playing a generative role with regard to variables. Yet, also as with flotation in liquids, we have a situation where the role of mechanisms cannot be all-encompassing. There are many variables, some relatively consensual, which cannot be generated by mechanisms. Furthermore, we have the maverick role of wind, a variable which appears to generate a mechanism. All in all then, we have a situation where conceptual structure as regards flotation in gases is roughly similar to conceptual structure as regards flotation in liquids.

How, then, about conceptual content? Here too, there are some similarities. Liquid stillness and movement were favourite variables with the airbed and plants scenes, and now we have no wind and wind featuring prominently with the balloons and magic carpet. Studies which have focused on properties of objects have shown that lightness and heaviness are further variables which cross the liquid–gas divide. And moving to mechanisms, the parallels continue. Flotation is explained in terms of downthrust and upthrust no matter what fluid we are talking about. Moreover, it is possible that when children exclude barriers from their accounts of gases, it is because gases are seldom thought to be strong enough to impose barriers. It may not be because barrier mechanisms are denied in principle.

Without doubt then, there are similarities between flotation in liquids and flotation in gases. However, there are also differences, and it is important not to lose sight of them. Even though lightness and heaviness are used as variables in both contexts, they are merely the most popular of a wide range of object variables with flotation in liquids. With flotation in gases, they dominate everything. As regards fluid variables, there is enhanced relevancy with age for flotation in liquids. With flotation in gases, there are no improvements with age. Furthermore, even though upthrust featured as a mechanism for both liquids and gases, the details of the concept varied considerably, as did the age trends.

Noting the points of difference, we can perhaps return to a theme introduced at the start of the chapter and ask whether they are sufficient to align flotation in gases with vertical fall and not with flotation in liquids. The answer must, surely and self-evidently, be 'No'. It is impossible to compare the variables generated for the parachutes and leaning tower scenes, the vertical fall scenes discussed in part III, with those generated for the balloons and the magic carpet scenes. The former variables were predominently object variables, while the latter were restricted to fluid. However, it is perfectly legitimate to compare the mechanisms associated with the two pairs of scenes, and this is not suggestive of linkage. First, the parachutes and leaning tower scenes produced mechanisms which did not have equivalents with the balloons and magic carpet. As mentioned in part III, these included inherent properties of the objects together with gravity. Second, the sense of upthrust in the context of the parachutes and leaning tower scenes was relatable to the received notion of air pressure. It was a far cry from the ephemeral phenomenon used with the balloons and magic carpet scenes. Finally, mechanism understanding improved with age with the parachutes and leaning tower scenes, $p = 0.06$ for the former but $p<0.01$ for the latter. There were no signs of age trends with the balloons and the magic carpet scenes.

In view of all this (and despite their phenomenological and scientific association), flotation in gases does not seem to be a sub-area of vertical fall. It seems on the contrary to be a sub-area of a category which also includes flotation in liquids. Indeed, it could show the same categorial relation to flotation in liquids that phase change was argued in part II to show to temperature change. In other words, like phase change, flotation in gases could be a peripheral, poorly analysed member of the category. Flotation in liquids could by contrast resemble temperature change in being central. Certainly, this gloss would fit with the present data. However, assuming it to be correct, why should it occur? The phenomenological differences between flotation in liquids and flotation in gases have been stressed repeatedly. Moreover, as the chapter started by showing, one route to overcoming the phenomenological, awareness of the pressure–density relation as it pertains to liquids and gases, is not typically available to children. How then does flotation in the two types of fluid come to be related? One possibility is that it is analysed in both cases as an *end-state* of a unitary set of actions. Although speculative, the possibility has the advantage of being consistent with: (a) the apparent discounting of dynamism for flotation in gases; and (b) the seeming dissociation from the instigating motion for flotation in general.

Further research is clearly required here. Nevertheless, no matter what transpires, the key point is that object flotation has emerged as a topic area in everyday physics which is distinct from propelled motion. Moreover, in emerging as distinct, object flotation has also shown parallels with heat transfer. These parallels may include a common central–peripheral structure. This remains to be seen. However, the parallels definitely include two further characteristics of considerable significance – the facts that both object flotation and heat transfer show signs of theoretical structure from a very early age, and that neither show systematic progress over variable and mechanism content until some years later. As argued earlier, such evidence suggests an action- and not theory-based model of conceptual growth, and within the action-based framework something which is not Piaget nor Vygotsky but rather somewhere in between. The point has now been reached therefore where we need to think carefully about how the model might work. This will be the substance of the chapter to follow.

Part V Conclusion

9 An action-based theory of conceptual growth

The book began with two approaches to conceptual knowledge, one of which it termed 'theory-based' and the other of which it termed 'action-based'. From the theory-based perspective, children have an a priori sense of mechanism which governs the conceptual distinctions that they subsequently draw. Driven by their mechanistic focus, children seek to understand how mechanisms work, and as understanding grows so does conceptualisation. This applies to all conceptual knowledge including beliefs about the variables relevant to outcome. Thus, the theory-based perspective makes the strong prediction that beliefs about variables will be governed by mechanisms, and hence that the anticipated progress in the latter will feed through to the former. From the action-based perspective, children have an a priori sense of personal action which they gradually decentre during the first six years of life. The course of decentration guarantees an awareness of variables prior to mechanisms, even though the early variables will be egocentric in nature. Moreover, the acquisition of a sense of mechanism around the age of 6 does not necessarily bring variables under mechanistic control. On the contrary, there is a lengthy period where some variables at least lie beyond the scope of mechanisms.

As things have turned out, the evidence presented in the preceding chapters has offered little support for the theory-based perspective. It is true that mechanisms were generative from an early age with heat transfer and probably object flotation. However mechanisms were never generative with the variables relevant to overall speed, and the implication was indeed atheoretical structure into adult life. Moreover even when mechanisms were generative, there was unmistakable evidence of variables beyond their scope. For instance, one study unearthed over 200 variables which 8- to 12-year-olds see as relevant to object flotation. Only a handful of these variables were generated by mechanisms. The action-based perspective can by contrast accommodate all these findings, and actually obtains more subtle support from what has gone before. For instance, the discussion of object flotation brought in genuine instances of egocentric variables in what it referred to as Level I responses. These responses were not very frequent but, when even 10-year-olds referred to bravery, concentration and effort as explanations of why objects float, completely undeniable. As noted already, egocentrism is predicted by the action-based perspective. However, it is hard to accommodate into the mechanism-driven development posited by the theory-based model.

Nevertheless, while the data pointed towards the action-based perspective, they did not unambiguously endorse either the Piagetian or the Vygotskyan line over fleshing the perspective out. The evidence that the heat transfer and object flotation mechanisms are generative from an early age does not sit comfortably with Piagetian theorising. While Vygotskyan theorising can cope with this, it cannot cope simultaneously with the fact that understanding of mechanisms and variables does not immediately improve. Yet delayed improvement was a robust finding across the previous chapters. With heat transfer and object flotation, there was little progress over either mechanisms or variables until 10, 11 and 12. With propelled motion, there was little progress until 14 or 15. Faced with such findings, a hybrid version of the action-based approach was proposed, neither Piagetian nor Vygotskyan but something in between. This version took from Vygotskyan theorising the idea that decentration is a question of co-ordinating actions with cultural symbols (and in practice with language). However, it saw the co-ordinations as less enveloping than Vygotsky proposed, such that children could try the ideas out and receive feedback. The period of zero progress comes about because children need to develop further before they can use feedback in an appropriate fashion.

It is difficult to anticipate how convincing the hybrid model will seem. Certainly, the preceding chapters should have made a fairly strong case against the theory-based perspective, and hence in favour of the action-based one. However, there may be a feeling that the proposed version of action-based theorising goes beyond what has been conclusively shown. If there are problems, they are most likely to lie with the claims about mechanism generativity. Without doubt, too much reliance was placed here upon single-observation 'snapshots', where the analysis was necessarily restricted to patterns of association. Gaps in the literature mean that longitudinal data of any kind were the exception rather than the rule. Particularly needed (and hence particularly valuable when they occurred) are longitudinal studies which use strategic interventions to expose dependencies. Nevertheless, against this, two positive points are worthy of note. First, there was never any inconsistency within topic areas. Analyses based on association patterns always concurred with each other, and were always in agreement with longitudinal data. Second, the variation between topic areas occurred despite essentially equivalent data-collecting procedures. This equivalence was taken to extremes within my own work, for there several topic areas were sometimes explored in the course of one study.

Obviously, the limitations in the data must be recognised, and indeed must be addressed through future research. Yet despite this, the positive points appear sufficiently compelling to treat the data and the model they support as indicative. Hence, they seem to recommend a more careful analysis of the model itself, how it might work in practice and what its workings are likely to mean. This is the rationale for the chapters to follow, with the workings of the model discussed in the present chapter and their wider implications in its successor. In relation to the workings, two issues will be focused upon: (a) how actions are co-ordinated with linguistic representations in contexts of relevance to everyday physics; and (b) how the content of representations is 'lost' in subsequent usage. Nothing will be said at this point about the origins of the egocentric representations which require co-ordinations in order to progress, nor about how the lost content will be found again once children attain a possibilistic perspective. By virtue of its action basis, the model is committed to the Piagetian account here, and this account has been outlined in earlier chapters. Thus, it does not need repeating in order to explain how the model might work.

CO-ORDINATIONS WITH LINGUISTIC REPRESENTATIONS

The focus of the present section is the co-ordinations which children might make between their egocentric knowledge and everyday physics as represented in language. As noted throughout, the concept of co-ordination between egocentric knowledge and linguistic representation is essentially Vygotskyan. Thus, the section will begin with a brief review of Vygotskyan writings, starting with Vygotsky's own work, to assess the relevance to physics. As it will turn out, the writings will prove to have limited relevance for they presuppose co-ordinations between actions and discourse *topics*. Everyday physics, it will be argued, seldom has topic status, but rather is called upon to support discourse around other topics. Recognising this, the remainder of the section will develop an alternative account which treats this supportive function as the stimulus to growth but which sees children as piecing everyday physics together in a largely unaided fashion.

Vygotskyan ideas on support to conceptual growth

From the translations of Vygotsky's writings that have now appeared (for example, Vygotsky, 1962, 1978), it is obvious that Vygotsky held very clear views about the role of cultural symbols in conceptual growth. These views stemmed from Vygotsky's belief that no matter what a given child is capable of individually in terms of received socio-cultural knowledge, that child can always do better in collaboration with skilled practitioners. Vygotsky used the phrase 'zone of proximal development' to refer to the gap between individual capacity and the potential for collaborative performance, and he believed that children vary markedly in the size of their zones.[1] However, regardless of size, the existence of zones means that there will always be opportunities to take children forwards in collaborative contexts, and for Vygotsky, these were opportunities not to be missed. They were for him nothing less than the prerequisites for growth, for he stated 'Every function in the child's cultural development appears twice: first, on the social level, and later, on the individual level; first between people (interpsychological) and then inside the child (intrapsychological). This applies equally to voluntary attention, to logical memory, and to the formation of concepts. All higher functions originate as actual relations between human individuals' (Vygotsky, 1978: 57). From this, it is obvious how Vygotsky saw the role of cultural symbols. It was to provide bridges whereby children could collaborate in actions which cross their personal zones. In Vygotsky (1962) he suggests that this may involve little more than 'the first step in a solution, a leading question, or some other form of help' (Vygotsky, 1962: 103).

This is interesting but exceedingly vague, and over the years there have been numerous attempts to become more specific. One is associated with James Wertsch (see Wertsch, 1985). Wertsch conducted a series of studies where mothers were instructed to work on a puzzle with their preschool children. A completed version of the puzzle was available, and the task was to copy this from a set of pieces. Wertsch was interested in the mothers' teaching strategies, in particular how they led their children to complete the puzzle and to internalise from this. As regards the latter, Wertsch stressed 'intersubjectivity', that is shared definitions of what the task involved. He believed that negotiation of reference was at the heart of intersubjectivity, both concrete negotiation at the level of labelling the pieces and more abstract negotiation regarding how the steps should be defined. On both counts there is close approximation to the ideas expressed by Wood *et al.* (1976). Wood *et al.*'s interest was in how mothers 'scaffold' tasks to maximise the chance of mastery. They identified six important tutorial functions, and two relate closely to Wertsch's: recruiting the child's interest in the task as defined by the mother and maintaining the pursuit of the goal through motivation and direction.

While not doubting for a moment that intersubjectivity and scaffolding are facilitative of learning, Rogoff (1990) reminds us that there is something both Western and middle class about the intensely co-ordinated activity that both imply. In other cultures and, perhaps, in other classes within our own culture, there is less emphasis on direct teaching and as a consequence fewer opportunities are provided for the explicit convergence of meaning. However in these other cultures/classes, there is, as Rogoff points out, an expectation that children will observe and participate in the everyday activities of adults. Rogoff uses the example of the Mayan people of Guatemala, who involve girls in weaving 'apprenticeship' from a very early age. This involvement is not, of course, silent, but nevertheless the instructional element is tacit and procedural. The 'explicit, declarative statements' (Rogoff, 1990: 119) implied by meaning negotiation do not occur. Rogoff's recognition of this leads her to subsume intersubjectivity and scaffolding within the broader concept of 'guided participation', a concept which emphasises 'the collaborative processes of (1) building bridges from children's present understanding and skills, and (2) arranging and structuring children's participation in activities, with dynamic shifts over development in children's responsibilities' (Rogoff, 1990: 8), but which does not entail direct negotiation of reference.

Because reference need not be negotiated, there will not necessarily be a collaborative product for children to internalise. Rather, there may be more of an adjustment to what Rogoff calls the 'social sense of their partners' (Rogoff, 1990: 196). As Rogoff points out, this is a case not of internalising the previously external but of moving gradually to the received socio-cultural wisdom by co-ordinating the presupposed with the new. There is, as a consequence, considerable emphasis on the active, sense-making efforts of children, efforts which Newman *et al.* (1989) have referred to as involving 'appropriation'. However, to return to our primary concern, what needs to be done in social interaction to ensure that appropriation will move children forwards? It has to be acknowledged that neither Rogoff nor Newman *et al.* offer clear guidance here. Indeed Newman *et al.* stress the importance of encounters where adults presuppose more common ground than actually exists, an approach which cries out for tighter specification. On certain readings, this could after all undermine completely the notions of intersubjectivity and adjustment to zones of proximal development.

Everyday physics and persuasive argument

Appropriation is vague but from another perspective it may also be too tight. One way in which Rogoff and Newman *et al.* continue the Vygotskyan line is by assuming that children will only appropriate from what is focal to the collaboration. No matter whether it is American children completing puzzles or Mayan children learning weaving, it is the focus of joint activity or discourse which is seized on and which effects the shift to socio-cultural wisdom. However, do children in the ordinary course of events make everyday physics the focus of attention? I think not, and ironically the very first sentences of Rogoff's book provide a perfect illustration of why I say this. These sentences describe a domestic scene in which 'Little David, at 7 months, sits in his rolling walker in the kitchen, as his mother cooks. His big sister Luisa, age three-and-a-half, sits on the floor beside him, telling him "Stay away from the stove . . . Hot! . . . Hot!" ' (Rogoff, 1990: 3).

In Rogoff's example, the focus of attention both in behaviour and, via the topic of conversation, in discourse is David's action towards the stove. Predicated on the topic is the assertion that he should stay away. Nowhere in the focus does everyday physics appear. However, the assertion is supported by a justification: David should stay away because the stove is hot, and this does contain an everyday physics message. Despite Luisa's relatively primitive syntax,[2] she is drawing attention to the fact that, despite not being inherently hot, stoves can heat up. However, her physics lesson is not (and does not become) the topic of conversation and hence cannot be said to be focal.

I think that what we have observed with Luisa and David may well be typical. In other words, although everyday physics is represented in discourse, it seldom becomes the topic of what that discourse is about. It is used instead to justify what is predicated on topics, as with 'Keep the door shut to stop the heat going out', and 'If you want those armbands to help, you'll have to blow them up properly.' In principle, there is nothing to prevent justifications from being adopted as topics as the conversation progresses, for example 'But I'm too hot; I want the heat to go out'. However, this kind of development strikes me as rare, particularly I feel when young children are involved. Doctoral research by Ed Baines (reported in brief by Baines and Howe, 1995) offers support here. The topic mainte-nance skills of 4- to 9-year-olds were documented in a number of task contexts with consistent results: discussion of justifications is exceedingly rare. The results are all the more powerful when it is noted that while some of Baines' contexts were 'naturalistic', others were contrived to make discus-sion of justifications a demand characteristic.

Suppose, then, that we accept that everyday physics is seldom taken as the topic of conversation. Where does this leave us as regards linguistic repre-sentations as a decentring force? If they are non-focal, linguistic representations cannot support the collaborative crossing of zones of prox-imal development, nor intersubjectivity through the negotiation of meaning, nor even guided participation with skilled practitioners. Can linguistic repre-sentations play any role therefore in the decentring of knowledge? The answer has to be 'Yes' and learning in a domain apart from physics, first language acquisition, shows us why this must be. Insofar as children learn the semantic content of their native language, they must: (a) decentre to the adult system; and (b) refer to linguistic representations (since there is no other source for what is itself linguistic). However, the primary route to semantic knowledge must be appropriation from non-focal experiences.

To see why non-focal experiences must be central to semantic growth, consider what may be the most transparent aspect of semantic content, word reference. It is true that parents frequently engage small children in conversations along the lines of 'What's that?', 'Ball', 'It's not a ball. That's a ball. This is an apple.' Such conversations make word reference focal, and they are undoubtedly useful to learning. However, they cannot be the only or even the main route to learning. Given Clark's (1993) observation that children average around 14,000 words by the age of 6, it is obvious that there are too many words to be mastered for them all to be learned in such an explicit fashion. In any event a large number of words lack the percep-tible referents which the conversational style presupposes. For both reasons, word reference must, in many cases, be pieced together from interactions in which it is not the focal concern.

Once we recognise the situation for language learning, we must acknowl-edge that none of the Vygotskyan mechanisms outlined in the previous paragraphs is essential for the co-ordination of egocentric knowledge with language. These mechanisms may be sufficient, but they are certainly not necessary. What then is required? I should like to suggest language serving communicative functions which, from the child's *egocentric* perspective, look as if they will be useful. Viewed like this, it is easy to see why linguistic representations should be relevant to physics. The point made above is that everyday physics is typically referred to in discourse to justify assertions. As such, justifications have an essentially persuasive value, that is they are used to strengthen the addressee's acceptance that the speaker's asserted beliefs or desires should be followed. However, persuasion is an essentially egocentric function: persuaders want to bring addressees into line with themselves. Thus, if the above proposal is correct and language needs paradoxically to serve egocentric ends in order to decentre, then language about everyday physics should neatly fit the bill. In doing so, it should fulfil precisely the role that our overall model demands.

PRESERVING STRUCTURE AND LOSING CONTENT

From the above, we can perhaps understand how linguistic representations could exert a decentring influence within everyday physics, even when the representations are not consistent with any Vygotskyan mechanisms. However, we have not so far considered how children could manage to preserve the structure of the representations but lose their content. Yet if the evidence of the previous chapters is correct, this is precisely what must happen between the ages of 6 and 11. In reality though, the depiction of everyday physics as used to justify in the service of persuasion renders the preservation of structure straightforward and the loss of content intelligible. Explaining why this is will be the main aim of the section to follow.

The processing of linguistic representations

Taking the preservation of structure first, consider a set of remarks which, if all that has gone before is correct, should typify what children hear about heat transfer and object flotation: 'Fasten your jacket to keep the cold out', 'Spread the burgers out so they'll defrost quicker', 'Relax so the water can push you up' and 'Hold the balloon tightly, so the wind can't carry it up.' Such remarks should be typical partly because they place everyday physics in a justificatory role and partly because they respect the generative role of mechanisms which has been hypothesised for the two topic areas. However, because justificatory remarks do not merely trigger decentration but also model communicative practices which children have incentives to adopt, both the discourse function and the mechanism–variable relation will be taken on board. Moreover both will be practised repeatedly as children try justification for themselves in the interests of persuasion.

An equally straightforward situation will apply with propelled motion, except that here the justificatory structures which are modelled and assimilated will fail to display mechanisms as generative. 'Try that bit of ice if you really want some speed' or 'Put on some top spin to slow it down' are more likely here. However, modelling and assimilation will continue to be apposite, which brings us face-to-face with the thornier question. If the structure of justificatory remarks is assimilated, the content must be assimilated also: so how is it that the variables and mechanisms which children propose up to the age of 11 are more primitive than those they propose later and hence (we assume) more primitive than the linguistic representations which they initially internalise? The answer lies, I think, with the justificatory function that the variables and mechanisms will serve in children's speech. Used to justify, variables and mechanisms will seldom be evaluated directly in the course of conversation. Any feedback which is forthcoming will be directed at the assertions being justified, and not at the justifications. Since children are relatively low in persuasive power (particularly in conversation with adults), the feedback on their assertions will frequently be negative, and the negativity will be carried through to the justifications. Either the whole message will be regarded as a communicative failure or the justification will be blamed as insufficiently compelling. No matter, the variables and/or mechanisms which children happily assimilated from mature everyday physicists will be called into question.

What will happen next is a rather moot point. Some might argue for the instant abandonment of the unsuccessful notions. I myself think that a more plausible account lies in the now popular 'connectionist' approach to conceptual organisation (see Plunkett (1995) and McClelland *et al.* (1986) for introductory reviews). This approach presupposes biologically given propensities: (a) to assign weights to experiences depending on their frequency of occurrence; and (b) to preserve associations between experiences via interconnections within neural networks. The first propensity means that if, say, lightness helps floating is frequently experienced as successful for communicative ends, it will acquire a heavy weighting. If vitamins help floating is not so experienced, it will acquire a low weighting. The second propensity means that messages which co-occur with messages about lightness and vitamins (perhaps because the same person says them) will be affected by the latters' weightings. Correspondingly, the weights attached to lightness and vitamins will also be affected by the weighting of connected ideas. The weights associated with the different ideas will determine their probability of being activated in appropriate contexts (for example, research interviews about everyday physics). Thus, the conceptual system at any point in time will be the set of the most heavily weighted ideas.

Thinking what this might imply for our present concerns, the crucial issue is how connectionism would deal with persuasive failures. To appreciate the options, let us imagine failures during the course of which lightness, airiness and bigness were referred to as helpful to flotation. By virtue of the failures, all three variables would receive negative weights. However, the three variables would be fed into a system where their weights would be readjusted depending on how many times they had been successfully or unsuccessfully mentioned in the past and how many times their associates had also been successfully or unsuccessfully mentioned. Let us suppose for now that lightness had previously been mentioned successfully on many occasions and never before unsuccessfully, airiness had never been mentioned before but was used this time in the context of lightness, and bigness although mentioned successfully many times before as a cause of sinking had never been mentioned before as helpful to floating.

Given what we are assuming regarding the past, the conceptual system to emerge from our three new experiences would be one where lightness would continue as a strong facilitator of floating (with its weighting marginally decreased), airiness would enter as a strong facilitator of floating due to its appropriation of the lightness weighting and despite its communicative failure, and bigness would continue as a strong facilitator of sinking (with its weighting unchanged or possibly increased). If bigness continued to be proclaimed as helpful to floating and these proclamations started to succeed, a point should come where bigness as a facilitator of floating and bigness as a facilitator of sinking should receive identical weightings, and the size dimension should in effect disappear from the system. Further proclamations of bigness as helpful to floating would lead to its being regarded as the facilitative factor.

The point about connectionism is, then, that although individual experiences are dealt with gradually and cumulatively, the immediate consequences for beliefs about variables can be negligible or dramatic, and consistent or reversing. It all depends on how experiences are embedded in the system of weighted associations. It is only after repeated and consistent experiences that trends will appear. Relating this back to children using everyday physics to justify, the implication is that over repeated cycles of negative feedback, the trend will be towards elimination, but the initial impact could be very different. With elimination though, children will have to cast around for alternative content, and there are essentially two strategies which they could follow. The first is to rescrutinise their linguistic experiences with, this time, an emphasis on content. The message from the previous chapters is that this will provide more options for variables than for mechanisms, and more options for object flotation variables than for any others. The second strategy is to induce variables and mechanisms from direct observations of relevant phenomena.

The success of the first strategy in providing accurate information will be directly proportional to the balance of relevant and irrelevant variables within everyday physics, and to the adequacy of the mechanisms which other people mention. Thus, we can refer back to earlier chapters to make predictions here. The success of the second strategy will depend on the accessibility of the variables and mechanisms to observation. In some circumstances, accessibility may be high: mention has already been made of the fact that heat transfer can be 'felt' directly. However, in most circumstances, accessibility will be low, and indeed rendered more problematic by two further considerations. First, ordinary life does not display variables and mechanisms in the ordered fashion of a science experiment, but rather confounds them. Thus, when massively confounded with metalness vs. woodenness, long cooking spoons might be more hazardous as regards scalding than short ones. Secondly, children are typically insensitive to confound. For instance, in research which involved 8- to 10-year-old children picking observations which would confirm or disconfirm hypothesised relations, Kuhn *et al.* (1995) found that only one child in a sample of fourteen consistently anticipated confound.

Probably then children will have limited success in finding adequate replacements for the variables and mechanisms which they come to eliminate. However, in one sense, this is not really the point. Once children obtain substitute variables and mechanisms, they will feed these into the communicative process, and have the same experiences of failed persuasion as they had with the original ones. Moreover, the next round of substitutions is just as likely with young children to shift back to the original ideas as it is to try something new: within the action-based framework, children do not have the capacity to keep track of experiences before the age of 11. This is not to say, of course, that the knowledge systems of 6- to 11-year-olds will all be equally bad. On the contrary, there will be variation between children at given points of time and variation within children as a function of time. The point is that progress will not be *systematic* before the age of 11, but will depend on which variables and mechanisms are most heavily weighted in knowledge at given moments in time. As time passes and weightings shift, there is as likely to be regression in individual cases as there is to be advance.

Computational and associative processes

The application of connectionism to everyday physics is not unique to this book. Proposals along these lines have already appeared in a lengthy article by diSessa (1993). Unlike this book, however, diSessa's article is not primarily concerned with the acquisition of everyday physics by relatively young children. Rather, its emphasis is the nature of everyday physics in comparatively sophisticated users. Thus, its main database is interviews conducted with approximately twenty students who were taking an elementary physics class at Massachusetts Institute of Technology. Nevertheless, the ideas voiced by the students will be familiar from previous chapters, for they included notions to the effect that objects need force to set them in motion, that once in motion objects acquire impetus, and that without continuous push impetus will dissipate. DiSessa's central claim about the students' notions was mentioned in passing during part I, namely that these notions can be characterised as 'phenomenological-primitives' (or p-prims). The notions are phenomenological in that they are superficial and readily accessible. They are primitive in that they are self-explanatory and atomistic. Being atomistic, p-prims are seen by diSessa as the elements of the knowledge system, and this is where connectionism comes in. Using what he describes as 'a fairly generic connectionist system' (diSessa, 1993: 180), diSessa claims that 'p-prims[3] have an external level, in addition to their internal one, that is supplied by connections to other p-prims; the external activation should be something like the sum of the (external) activations of all connected elements weighted by positive (activating) or negative (suppressive) factors' (diSessa, 1993: 180).

Clearly, diSessa's proposals have much in common with the line taken in the last few paragraphs. Nevertheless diSessa makes two claims which amount to crucial differences. First, he sees the whole of everyday physics as founded on a p-prim network that is structured in accordance with activation strengths. Second, he explains developmental change via: (a) the clustering of p-prims due to activation strengths; and (b) the abstraction of general principles across p-prim clusters. On present evidence, the first claim is too general and the second is wrong, and thus neither have equivalents in the preceding text. The trouble with diSessa's first claim is that although a p-prim basis could be argued from our propelled motion data, it does not square with what we observed for heat transfer and object flotation. Our key conclusion regarding heat transfer and object flotation has been that knowledge is theoretically structured from the very beginning. We could represent the elements in the theoretical structure, the mechanisms and attendant variables, as p-prims, but we would lose something crucial if we represented the relations between mechanisms and variables simply in terms of activation strengths. As has been stressed repeatedly, theoretical structure implies a generative relation between mechanisms and variables, and this cannot be captured via a frequency-based measure.

It is perhaps instructive that diSessa did not himself collect data relating to heat transfer or object flotation. On the grounds that 'intuitive knowledge is relatively rich in the vicinity of Newton's Laws' (diSessa, 1993: 125), his focus was exclusively upon force and motion. In view of what has gone before, it might be legitimate to suggest that this is an overly restricted database. However while this may be so, diSessa's model is lacking something even within the confines of force and motion. This brings us to the second claim and the second problem, namely the inadequacy of explaining developmental change in terms of p-prim clusters and abstracted principles. Quite apart once more from the fact that theoretical structure is basic with heat transfer and object flotation and hence not abstracted, clustering and abstraction could not possibly account for the step-wise nature of development observed with all three of our topic areas. Such development must reflect structural change external to the p-prim system, and we have been taking the parsimonious route of calling on structural change which is intrinsic to the action-based model. Kuhn *et al.* (1995) also recognise the need for structural change but refer to 'metastrategic competence', a disembodied entity which is unnecessary from the action-based perspective and not warranted by any data which Kuhn *et al.* present.

Up to a point, the line taken here is reminiscent of claims made by Lachter and Bever (1988) in relation to language acquisition and language processing. Having commented at length on the strengths and weaknesses of connectionist models, they propose 'a view of the mind as utilising two sorts of processes, computational and associative: the computational component represents the structure of behaviour, the associative component represents the direct activation of behaviors which accumulates with practice' (Lachter and Bever, 1988: 240). Nevertheless, despite the parallels, there are a couple of differences between Lachter and Bever's model and the present one. In the first place, Lachter and Bever refrain from claims about development while the present model treats associative processing as developmentally prior to computational. In addition, Lachter and Bever regard both associative and computational processing as progressive. They may be right as regards language, but the point made here is that associative processing is non-progressive in the context of everyday physics. It may lead to cognitive change during the period from 6 to 11 (perhaps even frenetic cognitive change) but the net effect of that change is the marking of time and not the boosting of knowledge.

Summary of the model

This then is what is being proposed: an action-based model which assumes: (a) the accumulation of egocentric variables during the preschool years; (b) the influence of justificatory remarks to produce decentred variables, a sense of mechanism, and in some cases a generative relation between mechanisms and variables; (c) the use of linguistic experiences in the production of justifications, and the experience of persuasive failure; (d) the refinement of mechanism and variable content in an initially unproductive (and possibly connectionist) fashion, but the preservation of mechanism and variable relations due to the structure of justificatory discourse; and (e) the acquisition of a possibilistic perspective around the age of 11, which allows the co-ordination of feedback on mechanisms and variables and hence their synthesis and gradual improvement.

As stressed repeatedly, the model is not particularly innovative. Quite apart from being action-based, it borrows heavily from both the Piagetian and Vygotskyan traditions. Nevertheless, it achieves a synthesis between Piaget and Vygotsky which is being increasingly argued for (see also Howe, 1993b) but which has not hitherto been attempted. Moreover, it is also markedly different from the theory-based approaches with which the book began and which have now been roundly rejected. In view of the rejection it is perhaps appropriate at this point to recall one of the reasons why the theory-based approaches occupied such a prominent and early position within the book: the approaches had straightforward and appealing implications for educational practice. Have these implications now been lost? The chapter to follow will start by attempting to answer this question. However, as it proceeds it will unearth further issues of psychological as well as educational significance. It is with these issues that the chapter, and hence the book, will end.

10 Action-based knowledge in a wider context

The intention for this chapter is to discuss the wider implications of the action-based model we have ended by endorsing, starting with the consequences for educational practice. More specifically, the aim is to consider our model against the theory-based approach which we have now rejected, but which was freely acknowledged to have attractive educational implications. With this in mind, we need to remind ourselves of why the theory-based approach seemed educationally so appealing, and as signalled in part I, there are probably two main reasons. First, the implications for teaching are entirely general: they do not imply differing practices as a function of topic area. Second, the implications are for a focus on mechanisms with variables being seen to take care of themselves.

It is obvious that our action-based model does not have the second of the above two advantages, but then the evidence indicates that the advantage is mythical. As predicted by the action-based model, there are variables beyond the scope of mechanisms even when the latter are generative. In addition, there are topic areas where whole clusters of variables are unrelated to mechanisms. However, what about the first advantage of the theory-based approach, the fact that its educational implications are general across topic areas? We have observed differences between all three of our topic areas, differences over the relative orthodoxy of mechanisms, over the extent of their generativity, and over the variables which lie beyond them. The age trends also showed some variation as a function of topic area. Our action-based model can account for these differences, so does this mean that as regards teaching it implies contrasting strategies for each distinct topic area? This is the first question to be addressed in the chapter. The message will be that although teaching strategies will have to respect topic area idiosyncrasies to some extent, general principles can also be established.

However while the establishment of principles with generality across topic areas has to be welcomed, there is another angle to the issue of generalisation which also requires discussion. Our action-based model predicts changes with age in the structure of knowledge, and these predictions have been amply supported in earlier chapters. Perhaps though these age changes have implications in their own right for the practice of teaching. Perhaps indeed, they mean contrasting strategies as a function of age, compromising the partial generality across topic areas with a specificity across pupil cohorts. This is the second question to be discussed in the chapter, with the conclusions turning out to be more optimistic than might currently seem likely. However, in reaching these conclusions, the chapter will also highlight issues of psychological significance. Foremost amongst these is the resurrection (and particular interpretation) of the concept of cognitive 'stage'. Thus, it is with psychological theory that the chapter will end.

SENSITIVITY TO TOPIC AREA

To begin then, the present section will focus on the issue of how sensitive to topic area teaching should be. It will start by introducing a group of psychologists who would not be in the least surprised to find sensitivity proving essential. These psychologists have become persuaded that conceptual knowledge is encapsulated into identifiable 'domains', such that 'the manner in which information is processed in one domain may be different from the manner in which it is processed in another' (Maratsos, 1992: 3). Since differential information processing seems to necessitate differential information provision, topic-sensitive teaching appears inevitable. Up to a point, this will be conceded. Nevertheless, it will be argued that what applies to information provision does not necessarily apply to information use. Moreover, given what has already been claimed about everyday ideas, use of received concepts by pupils is a sine qua non for effective learning. Recognising this, the section will end by identifying a set of teaching practices which should have applicability regardless of topic.

The relevance of domain specificity

The psychologists who advocate conceptual domains usually trace their intellectual roots back to Chomsky's (for example 1965, 1981) claim that linguistic knowledge is organised into autonomous modules. However, they are often also heavily influenced by Fodor's (1983) extension of Chomsky's ideas to the 'input systems', that is language plus the perceptual processes. Nevertheless, as exemplified by Carey (1985b), Karmiloff-Smith (1992) and Keil (1986), the notion of domain-specific conceptual knowledge goes beyond Chomsky and Fodor, and indeed asserts something which Fodor explicitly denied. Conceptual knowledge is located in what Fodor referred to as the 'central systems', and these systems were for Fodor necessarily unencapsulated.

The question which we need to consider is whether in the light of what we have observed for heat transfer, propelled motion and object flotation, we should follow Fodor or Carey and co. In other words, we must decide whether our observations suggest that conceptual knowledge in the three topic areas is or is not organised into domains with information-processing implications. By way of an answer, the first point to note is that the contrasts between the topic areas were not themselves the products of information-processing differences. The processes listed at the end of the previous chapter as characteristic of the action-based model were, without exception, domain-general. The topic area differences were seen to come about because of the interaction of the processes with two kinds of input: (a) sensori-motor knowledge, which renders some physical phenomena (for instance, the 'feel' of heat being transferred) more accessible than others; and (b) cultural-symbolic representations, which contrast between topic areas in both structure and content.

Noting that the differences originate despite common processes rather than because of them, it is unclear whether Maratsos (1992) would accept that our topic areas were genuine domains, for he believes that 'domain specificity means there is something *innate* in the organism that causes responses in that domain to be different in some way' (Maratsos, 1992: 3). It is true that sensori-motor knowledge is, on the action-based model, trace-able to the innate schemes of the neonate, but this is a far cry from claiming that the distinction between, say, propelled motion and object flotation is innately specified. On the other hand though, Keil (1990) is happy with the idea of acquired domains, and yet, as noted already, he subscribed to the view that domains have information-processing implications.

Keil's view is in fact perfectly tenable. It is quite possible that domains emerge under the operation of common processes, and yet once they have emerged require different strategies for subsequent growth. Up to a point, I think that this could be the case for the teaching of physics. From everything that has gone before, heat transfer probably could be approached in a fashion not dissimilar to what the theory-based perspective would advocate. In other words, mechanisms ought to be the focus, and it could be assumed that children would treat these mechanisms as generative of variables. On the other hand, children would probably still need some guidance over appropriate variables, and they would definitely need to be told that the *only* variables to be considered were the ones that stemmed from mechanisms.

With propelled motion, the approach would almost certainly have to be more indirect, and one possibility would be to start with speed change, that is acceleration and deceleration. This is because, as part III demonstrated, children hold theories about how change is effected. Thus, it might be possible to exploit their theories to introduce Newton's Second Law, and from this the implications of friction and drag. Since, as we saw, these impli-cations make reference to both velocity and variables, this could be the route to simultaneously linking speed with speed change and mechanisms with variables. It could also be the route to refining the variables which children's everyday physics treats as significant into something more appropriate.

The point is that the way in which children's knowledge becomes struc-tured probably does have implications for how the received wisdom should be packaged. Interestingly, the implications for propelled motion sketched in the previous paragraph are the reverse of standard teaching practice. They start with acceleration and deceleration and introduce velocity derivatively when, following Newton, educational orthodoxy is to do the reverse. This aside, packaging obviously is important, and it should seemingly vary with topic area. However, it is almost certainly not the only thing which needs to be considered and, once we recognise this, the force of the emerging domain specificity is somewhat reduced.

In particular, if the line taken in the previous chapter is correct, we must assume that children's everyday physics at any given moment is the sum total of the arguments which they currently regard as persuasive. Thus, to improve children's understanding, it cannot be sufficient to communicate received beliefs, no matter how appropriately these are packaged. It must also be necessary to put children in situations where they experience received beliefs as persuasively powerful (and, at the same time, everyday beliefs as persuasively weak). Otherwise, the message of schooling will be lost in the mêlée of ordinary communicative practice. However, experiencing persuasive power and weakness is a question of being party to certain forms of social interaction, and has nothing to do with knowledge domains.

Persuasive argument in the physics classroom

Arguably then, the job for teachers is to present physics 'truths' in a fashion which respects the structure of everyday ideas, and to organise classroom activities such that the truths are relatively persuasive. The presentational side would seem unproblematic given adequate information about how everyday ideas are structured, but what about the persuasive? Here, there would appear to be two issues in need of resolution: (a) how to ensure that children receive compelling evidence of persuasive power and weakness; and (b) how to guarantee that the evidence for power is aligned with acceptable ideas and the evidence for weakness with unacceptable.

Starting with the first issue, it seems obvious that contexts should be provided where children are themselves involved in the persuasive process. However, if this is the case, it is difficult to see how those contexts could involve social interaction with the classroom teacher. Children will almost invariably presume that their teacher knows more about physics than themselves. Thus, they will find something phoney about trying to persuade a teacher that his/her ideas are wrong while their own are correct. Moreover, they will treat persuasive overtures from the teacher as didactic rather than persuasive, persuasion implying more respect for their own perspective than they will assume to be warranted. Either way, teacher–child interaction is unlikely to work in the intended fashion, a point related interestingly enough to an observation made by Piaget (1932). Addressing moral reasoning rather than physics, Piaget commented that in any adult–child interaction the authority of adults undermines the argumentative stance of children.

For Piaget, the possibility of persuasion depended paradoxically on differing opinions within roughly symmetrical status. This led him to propose peer interaction as a more suitable forum for persuasive encounters, a proposal explored and supported in relation to morality by Kruger and Tomasello (1986 – see also Kruger, 1992). Thinking about physics, there is not, as far as I know, any work which has explicitly requested persuasion in peer interaction. However, the demand characteristics of a series of studies that my colleagues and I have conducted (Howe *et al.*, 1990, 1992a, 1992b, 1995b) have produced something equivalent.

The participants in my studies varied in age from 8 years to 18 and over, but the studies themselves had a constant three-stage structure. The first stage involved individual assessments (usually interview-based) to establish initial conceptions about issues relevant to physics. Since the issues related amongst other things to propelled motion and object flotation, the first-stage results have in some cases been reported in previous chapters. Not hitherto reported is the second stage where participants were asked: (a) to make written predictions about events, for example whether a metal box would float or sink, and whether a heavy vehicle would roll a short or long distance; (b) to share their predictions with other participants and come to an agreement as to what would happen; and (c) to test the agreed prediction empirically and reach a joint interpretation about what transpired.

Although the instructions stressed agreement, it was clear from video-recordings of the discussions that the participants had considerable ego-investment in their written predictions, and were keen to see them adopted as the collective position. Thus, when predictions differed, there were, as the following extract illustrates, unmistakable attempts to persuade. In the extract, four 10- to 11-year-old boys are discussing which of three squares on the floor will be reached when a gate on a slope is raised and a toy car rolls down and off. They had previously seen another toy car reach the middle square in conditions that were equivalent apart from the slope's surface:

ANDREW   It'll go to the same square as the other one.
SIRINDER I agree.
ABRAR   I think it'll go to the furthest square.
SIRINDER What makes you think that?
ABRAR   It's a good surface. It's slippy, and it's a good surface.
KEMAL   But it's the middle gate.
ANDREW   And the same weight.
SIRINDER We all think the same except you, Abrar. Can't you agree?
ABRAR   All right then, but it's still a smooth surface.
SIRINDER So we all agree the same square.

There is no doubt that conversations like the one illustrated were beneficial to the participants' understanding. The third stage in all of the studies involved reassessing the participants on an individual basis to establish their post-interaction beliefs. The progress from the initial assessments to these third-stage assessments was always significantly greater when the interactions were between participants with differing opinions than when they were between participants with similar opinions. When opinions were similar, persuasion was obviously unnecessary. In addition, dialogue analyses conducted as part of the studies reported by Howe *et al.* (1992a, 1995b) established a direct link between persuasive exchanges and post-interaction progress. Importantly, it did not appear to matter whether the participants undertook the role of persuader or persuaded, for the overall pattern of group interaction was as good a predictor of subsequent progress as the contributions made by individual participants.

Nevertheless, while the value of persuasion was clearly demonstrated, the magnitude of progress in the studies was limited. As intimated in earlier chapters, the first- and third-stage responses were scored on scales of between four and seven points. Although mean progress between the first and third stages was always statistically significant in the 'differing conditions', it never amounted to as much as one scale point. The reason for the limited progress may lie with the manifest failure of the studies to fulfil our second requirement, that persuasive power be aligned with acceptable ideas and persuasive weakness with unacceptable. The interaction between the four boys quoted earlier is an excellent illustration of this, for when the gate was raised and the car rolled down the slope, it reached the further square and not the jointly predicted middle one. This prompted the following reaction:

ALL   Oh no-oh!
ABRAR   I was right and I changed my mind. I was right.
SIRINDER   Well, we all make mistakes in this world.

For the boys, persuasive power was aligned with poor ideas, and persuasive weakness with good. What could have been done to reverse the situation? The unfortunate experience of Abrar suggests that the presence of participants with good understanding is not sufficient. Far from persuading the others, such participants can be swayed by a dissenting majority and/or a socially dominant opponent. However, perhaps Abrar's contribution would have been more forceful if there had been another group member with equivalent understanding. Certainly the extensive social psychological research into group conformity would lead us to predict this (see, for example, Allen, 1975). Unfortunately though, to the extent it was the case, it would appear to imply screening by teachers prior to grouping to establish whose understanding is relatively good. In addition, it is possible that a contribution such as Abrar's would have been more forceful if the group work had been embedded in a context of teacher presentation – persuasive activity – teacher presentation etc. It was beyond the remit of my studies to establish such contexts, but they are implied by the line taken in the present chapter. Certainly, the possibility of referring in groups to material mentioned by teachers on previous occasions ought to increase the probability of a received 'voice' coming through without decreasing the probability of genuine persuasiveness.

Besides ensuring that group activities are balanced by teacher presentations and that expert group members receive social support, it might also be desirable to make small modifications to the tasks which the groups carry out. In the studies which I have been describing the main persuasive effort went into the predictions. This comes out in the dialogues which were quoted above and it is not remotely surprising once we remember that the predictions were where the written commitments were made and the ego-investment correspondingly strong. However, the participants with the best ideas did not have confirmatory feedback to support them during the predictive phase, and thus may not have made the case which they could have made during the subsequent interpretive phase. Suppose, though, that the interpretive phase was given a boost to make it equally significant in the participants' eyes. This would give the 'good ideas' people extra chances to shine. One way to achieve the boost would be to build on the 'rule generation' strategy of Howe *et al.* (1995a) and Tolmie *et al.* (1993) that was reported in previous chapters, and ask for the interpretations to be agreed *and* written down. As well as giving the discussions 'closure', written explanations would provide (and be seen by participants to provide) externally verifiable records of group activity. Slavin (1983) has evidence that in its own right this is motivating and helpful.

Of course boosting the interpretive phase will only work to the extent that the feedback endorses the expert line. Thus, great care will have to be exerted over the events to be interpreted. Besides being consistent with received physics knowledge, the events should ideally also be inconsistent with as many unorthodox ideas as possible. Thus, in a task relating to the variables pertaining to flotation, it would seem advisable to present wooden or airy objects that sink and metal or solid objects that float. It would obviously be impossible to provide counters for all the unorthodox ideas that appear in everyday physics. We have noted repeatedly how the pupils in the Howe *et al.* (1990) study came with over 200 variables relating to flotation in liquids, and as will be remembered from part IV the majority of these were scientifically irrelevant. Rather, the point would be to provide events that were inconsistent with a good selection of the commoner irrelevancies, and the reason for doing this would be to provide evidence that could be fed into the discussions. In other words, it would allow an individual with good ideas to say, as one did in the Howe *et al.* (1990) study, 'I don't think that being wooden has anything to do with it. That one's wood and it sank. The important thing's how light it is for its size.'

To recap on what has been suggested so far, the key proposal is to build on the essentially persuasive orientation that children have towards everyday physics. This has been taken to imply science lessons where children participate in joint decision making. The decision making will be in the context of tasks where children have: (a) to decide together what the outcomes of events will be; (b) to test their decisions empirically and discuss why things turned out the way that they did; and (c) to pool their ideas in the interests of the best possible interpretations and to write these down also. Because of the incompatibility between persuasion and asymmetrical status, the decision making should only involve children. The classroom teacher should not participate. However, the classroom teacher will still have a crucial role to play. First s/he will have to present the information which the groups will in a sense 'practise' on a subsequent occasion. Second, the teacher will have to screen the children to establish who has relatively good understanding. Screening could involve paper-and-pencil versions of the interview techniques reported throughout the previous chapters. Third, s/he will have to ensure that the children with relatively good understanding are strategically distributed for the joint decision making. Fourth, the teacher will have to organise the empirical testing to maximise support for good ideas and minimise support for weak. Thus, the teacher's role will be central, albeit less direct than traditionally construed. However will it produce a workable package, and will it work for all children? Such issues will be discussed in the section to follow.

SENSITIVITY TO AGE OR STAGE

As noted on several occasions, physics is traditionally taught relatively late in the school curriculum, typically not before the teenage years. Within the past decade, there has however been a move to change things by introducing physics at a much earlier point, perhaps even during the first years of schooling. This results from a new take on the social problem with which the book began, the widescale abandonment of physics relative to other school subjects. It is thought that if physics is encountered early in schooling, its status amongst pupils will rise and its content can be covered at a more leisurely pace. This has led in the United Kingdom to detailed programmes for all age levels, in particular the National Curriculum for England and Wales (Department of Education and Science, 1989; Department for Education, 1995) and Environmental Studies 5–14 for Scotland (Scottish Office Education Department, 1993). The programmes have not been implemented as widely as hoped for at this point in time. Nevertheless, with them as background, it might be advisable to think about our recommendations for teaching with reference to all age levels rather than just teenagers. However, if we do this, are we not bound by the age differences documented in previous chapters to anticipate problems? This question will be the focus of the section to follow.

Receptivity to feedback

From a teaching perspective, the most worrying element of the model consolidated in the previous chapter (not to mention the data presented in its predecessors) is, surely, its depiction of young children as responding in a disorganised fashion to their everyday experiences. They are sensitive to feedback, to be sure. However, their knowledge changes in response to feed-back are unsystematic, meaning that progress is unsustainable and backtracking common. These tendencies were attributed to a cognitive limi-tation, namely the lack of a possibilistic perspective, but this means surely that the tendencies may have relevance to the educational settings which are now being discussed. If cognition makes young children unsystematic in their everyday lives, why should it not do the same in contexts of teaching? Yet if it does this, it would seem to imply that no matter how successful the interventions sketched in this chapter might prove to be in the short term, they could never with young children lead to durable progress. Such progress would be restricted to children aged 11 and over.

The idea of cognitive structures which differentiate between everyday situations and teaching is of course completely untenable. Nevertheless, there are grounds for thinking that if the techniques described in this chapter did achieve their short-term goals, the results could be robust against the buffeting effects of cognition. In other words, the techniques might protect young children against their 'natural' tendencies.[1] The grounds for thinking this are partly empirical. The third-stage assessments which featured in my work on peer interaction all took place some weeks after the group tasks. In some cases, the post-group interval was as much as eleven weeks. Far from showing decay, the third-stage performance was, if anything, better than that achieved during the group tasks. This was despite the strong links already mentioned between third-stage performance and group events. No matter how such results are to be interpreted, they suggest that peer interaction as sketched above can be an extremely powerful experi-ence, sufficient to carry children through their later social encounters. In addition though, a recurring theme of the book has been that the more chil-dren's ideas approximate scientific truths, the less susceptible they are to subsequent loss. This is regardless of children's age levels. In theory, the teaching techniques could take children to Newton's Second Law or Archimedes' Principle, and once this happened there should be considerable resistance to loss.

A 'stage-theoretic' finale

Note though what is being claimed. In accordance with the action-based framework, children below the age of 11 have cognitive structures which differ from older children and adults. However, experiences can be organised to shield them from the adverse consequences that these structures would normally have. Thus, despite their limitations, young children can master the relevant material, and they can do this in a durable fashion. In the parlance of developmental psychology, the claim is 'stage-theoretic', but (and this is crucial) stages are being theorised in a fashion which departs from tradi-tional practice. Traditionally, stages are said to place *constraints* on what can be learned, that is to preclude certain outcomes. Here the outcomes are attainable so long as the learning experiences are sensitive to the limits imposed by stage.

Nevertheless, being stage-theoretic (albeit unconventionally stage-theoretic), the present claim contrasts markedly with a current trend in developmental psychology, a trend which is well represented by Carey (1985b). Asking whether children are 'fundamentally different kinds of thinkers and learners than adults', Carey argues that in general the answer is 'no'. Domains are for Carey the major influence on styles of thinking and learning, and not cognitive maturity. However, the evidence which Carey uses to support her argument is studies showing that with highly structured tasks children as young as 3 can produce performances which are qualitatively similar to those produced by adults. The trouble is that stage in our sense permits similar outcomes, which after all is what the 'performances' discussed by Carey amount to. Moreover, it does this in circumstances where experiences are sufficiently structured to offer protection. Thus, the possibility is raised that the work cited by Carey achieved its effects by doing precisely what our putative teaching techniques are being claimed to do, shield children from themselves. Far from denying stages therefore, Carey's database precludes them from being displayed.

The implication is of course that to expose stages, should they exist, it is necessary to work with relatively unstructured tasks. Can such tasks be tolerated? I realise that by seeming to advocate these I am now skating on very thin ice. The history of developmental psychology can be depicted as a shift away from the unstructured interviews favoured by Piaget to the highly controlled investigations of contemporary scholars. There can indeed be little doubt that Piaget was guilty of serious methodological sins, but on balance I do want to back him here on the issue of structure. There is no denying that the relatively unstructured investigations described in previous chapters unearthed a richness to children's thinking about physics that was missed by their more constraining counterparts. Moreover, it was this richness which motivated the conclusions about cognition which are currently under discussion.

Thus, the book can perhaps end on a pro-Piagetian note, and this arguably is only right and proper. For all the limitations discussed in earlier chapters, Piaget has provided the most complete specification of the action-based framework, which the book has now endorsed. Moreover, the guidance that Piaget provided about age 'norms' has largely been confirmed. Piaget's theory is certainly inadequate to deal with all aspects of everyday physics, and we now have a new model which is an attempt to improve things. Nevertheless, the model lies within a framework that was laid down by Piaget and embellished by Vygotsky. This most definitely must not be forgotten.

Appendix – methodological details of the Howe (1991) study

The study was intended to document pupils' pre-instructional ideas relating to four topic areas of relevance to physics. The areas were heat transfer, object flotation, matter transformation and propelled motion. For each area, interview schedules were prepared tapping pupils' ideas about: (a) the outcomes of four key events; (b) the variables relevant to the facilitation or inhibition of outcomes; and (c) the mechanisms by which outcomes are achieved. The schedules were used to structure one-to-one conversations between the interviewer and each participating pupil.

PARTICIPANTS

A total of 126 pupils participated. They were all attending schools in a rural and predominantly middle-class area that was located within commuting distance of Glasgow, Scotland. The distribution of the pupils by school class, age and gender was: thirteen boys and thirteen girls from Primary Two who were aged 6;3 to 7;3 years (described as the 6- to 7-year-olds); twelve boys and thirteen girls from Primary Four who were aged 8;3 to 9;3 years (the 8- to 9-year-olds); twelve boys and twelve girls from Primary Six who were aged 10;3 to 11;3 years (the 10- to 11-year-olds); thirteen boys and eleven girls from Secondary One who were aged 12;3 to 13;3 years (the 12- to 13-year-olds); and fourteen boys and thirteen girls from Secondary Three who were aged, with one exception, 14;3 to 15;3 years (the 14- to 15-year-olds). The exception was a 16-year-old girl who, having recently arrived from overseas, was being obliged to follow the Secondary Three curriculum. Although gender was not a variable in the study, it seemed advisable to equalise as far as possible the numbers of boys and girls, for there is some (controversial) evidence for gender-related differences in physics knowledge (see, for example, Becker, 1989). The 12- to 13-year-olds and the 14- to 15-year-olds were recruited from the one secondary school in the area of the study, the younger pupils from two of its primary 'feeders'. The 12- to 13-year-olds had received teaching related to some of the topic areas covered in the interviews. The 14- to 15-year-olds had received teaching related to all of them though in some cases at an introductory level. None of the topic areas had, however, featured in the teaching to the three youngest age groups.

MATERIALS

The interview schedules addressed sixteen photographed scenes. Four scenes depicted events relating to the transfer of heat. These were four pans sitting on a cooker – one containing water (described as the pans scene); four forks around a lighted barbecue, one being used to cook a sausage (the forks scene); four saucers on a shelf above a radiator, one holding a plant (the saucers scene); and five wraps on a picnic rug, one being placed around ice cream (the wraps scene). Another four scenes depicted events relating to the flotation of objects. These were: a swimmer on a gradually sinking airbed (the airbed scene); a basket of plants floating around in a garden pond (the plants scene); three helium balloons in front of a tree prevented from floating by a heavy brick (the balloons scene); and two cartoon gnomes flying on a magic carpet (the magic carpet scene). A further four scenes depicted events where the appearance (and in two cases, chemical composition) of matter was being or could be materially altered. These scenes were: charcoal burning on a barbecue (the charcoal scene); four nails on a garden table, one being hammered in (the nails scene); three wax candles alight on a dining room table (the candles scene); and a block of jelly plus a spoon, knife, kettle, mixing bowl, measuring jug and chopping board, all laid out on a kitchen table (the jelly scene). The final four scenes related to propelled motion in a vertical or horizontal direction. These scenes were: a large green ball being rolled across paving stones as in hopscotch, with a tennis ball, table tennis ball, golf ball and bowls wood in the immediate vicinity (the hopscotch scene); an Action Man being lowered on a plastic parachute, with three alternative parachutes in the vicinity (the parachutes scene); the bowls wood being 'curled' across an ice rink, with the other four hopscotch balls in the vicinity (the curling scene); and the Leaning Tower of Pisa (the leaning tower scene).

The interview schedules comprised a string of questions for each of the scenes. These questions are shown below, though to save space a degree of editing has taken place. Text which helped the pupils contextualise the scenes and/or the major questions has been omitted. Questions phrased like 'What do you think is important?' and 'What makes you think this?' are abbreviated into cursory 'What?' and 'Why?', and the options for modifying questions according to pupil answers are not fully shown. Occasionally though, manifestly contingent wording is presented in brackets.

Pans scene   What will happen to the water once the cooker is switched on? How will that happen? What will happen to the water's temperature? Will the temperature go on (rising)? Does the kind of pan make a difference to how quickly the water will heat up? Has the cook chosen the best pan for heating the water quickly? Which pan would be best? Why? Why is (variable) important? Would the other pans be equally bad or would some be better than others? Why? Why is (variable) important?

Fork scene   Why is the cook holding the sausage with a fork? Do you think that her fingers could still burn even though she's got a fork? How will her fingers burn? Does the kind of fork make a difference to whether fingers will burn. Has the cook chosen the best fork to keep her fingers from burning? Why? Why is (variable) important? Would the other forks all be as bad as each other, or would some be better than others? Why? Why is (variable) important?

Saucers scene   Suppose you leave the saucer over the radiator. Will all the water be there after a couple of days? Where will it go? How does that happen? What will happen to the water's temperature while it is (evaporating)? Do you think that the kind of saucer makes any difference to how long the water will last? Do you think the water will last longest in the saucer that is under the plant? Why? Why is (variable) important? What about the other saucers? Will the water last about the same time in them, or will there be differences? Why? Why is (variable) important?

Wraps scene   What happens to ice cream when the sun shines on it? How does that happen? Do you think that wrapping something around ice cream will make any difference to how quickly it melts? What difference will it make? Do you think that some wraps are better than others at keeping ice cream from melting? Has the picnicker chosen the best wrap? Why? Why is (variable) important? What about the other things? Would they all be equally bad or would some be worse than others? Why? Why is (variable) important?

Airbed scene   What will happen to the swimmer if all the air comes out of the airbed? How will the water feel as she goes under? Will the water change in any way as she goes deeper and deeper? How? How does this happen? How could the airbed be made to float higher? How about doing something to the water? Why is (variable) important? Like, would it make any difference if the swimmer went shallower or deeper? Why? There's a jacuzzi next to the pool where the water is rougher. Would the rougher water make any difference? Why? Do you think anything could be added to the water to help the airbed to float? What? Why would that help?

Plants scene   How could the plants be made to stay under? Do you think that anything could be done to the water itself? What? Why is (variable) important? Would it make any difference if the basket was moved to a deeper or shallower part of the pond? Why? Suppose we put on the garden fountain to make the water rougher? Would this have an effect? Why? Could you add something to the water that would help? What? Why would that help?

Balloons scene   What will happen if we take the brick away? Imagine the balloons could lift a very tiny child. Will the sky feel differently to the child as she goes higher and higher? Why? Will the sky change in any way as she goes higher and higher? How? How does this happen? Let's think about the balloons without anyone holding on. Suppose that we had a competition to see which balloon went the highest? Which one would you like to win? Is there anything that you could do to help it? Why is (variable) important? Would it make any difference if you climbed the tree before letting go? Why? Or took the balloon up in an aeroplane? Why? Would it make any difference if the weather changed in some way? Why? What about going into Glasgow where the sky is dirtier? Would that make a difference? Why?

Magic carpet scene   The point about magic carpets is that they are supposed to fly without engines. Can you think of anything else that flies without an engine? (The preceding questions were omitted for the 6- to 7-year-olds and the 8- to 9-year-olds, and the next question addressed magic carpets explicitly.) Have you ever thought about how (such things) get down from the sky once they are up there? Tell me what you've thought? Why is (variable) important? Do you think that aiming towards hills where the ground was not so far below would make any difference? Why? Do you think that the kind of weather makes a difference? Why? Would it make any difference if the sky was clean or dirty? Why?

Charcoal scene   What is burning inside the barbecue? When the (charcoal) stops burning, what will be left? Are charcoal and (ash) made of the same stuff, or are they made of something different? How does charcoal change into ash? It's not usually a good idea to cook until the charcoal has changed into ash. Suppose you were hungry and wanted to speed the change up. Is there anything you could do? What? Why? Why is (variable) important? Would it make any difference if you poked the charcoal? Why? Do you think it matters where you stand the barbecue in the garden? Why?

Nails scene   Suppose we have a look at that nail after the table's been outside for a year. Do you think that it will still look the same? How will it have changed? Do you think that (rusty and non-rusty) nails are made of the same stuff, or are they made of something different? What makes rust form? Do you think that the kind of nail makes a difference to how quickly things rust? Do you think that the joiner has chosen the nail that will rust most slowly? Why? Why is (variable) important? How about the nails that wouldn't be so good? Would they all rust equally quickly or would some be slower than others? Why? Why is (variable) important?

Candles scene What will happen to the wax as the wick burns down? Are (melted or non-melted) candles made of the same stuff, or are they made of something different? What is happening to the wax while it is melting? Suppose that there was a power cut and you wanted the candles to last as long as possible. Is there anything that you could do to slow the melting down? What? Why? Why is (variable) important? Would putting something over the candles make any difference? Why? Would moving them to another part of the room make any difference? Why? Would it make any difference to make the room cooler? Why?

Jelly scene Do you know how to make a jelly? Is a jelly made of the same stuff before and after it's been (dissolved) in water? What is happening to the jelly when it is (dissolving)? Suppose that you were in a hurry and wanted the jelly to (dissolve) as quickly as possible. Is there anything you could do to speed things up? What? Why? Why is (variable) important? Would the amount of water make a difference? Why? Would stirring make a difference? Why? Would the size of the jug you put it into? Why? Would the size of piece? Why?

Hopscotch scene What will happen to the ball's speed as it rolls across the paving stones? Why? Do you think that, no matter how hard you push, some kinds of balls could always be made to roll faster across the paving stones than others? Why? Has the player chosen the ball that could be made to roll the fastest? Why? Why is (variable) important? Which ball could be made to roll the fastest? Why? Why is (variable) important? What about the balls that couldn't be made to go so fast? Would they all go equally slowly across the paving stones, or could some be made to go faster than others? Why? Why is (variable) important?

Parachutes scene What will happen to the parachute's speed as it falls through the sky? Why? The best parachute would be one that fell very slowly. Could anything be done to the parachute that the toy is holding to make it fall more slowly? What? Why? Why is (variable) important? Look at the other parachutes on the floor. Do you think that any of them could be used to make a better parachute? Which? Why? What about the others? Would they all be equally bad, or would some be better than others? Why? Why is (variable) important?

Curling scene What will happen to the ball's speed as it rolls across the ice rink? Why? Do you think that no matter how hard you push, some kinds of balls could always be made to roll faster across the ice than others? Why? Has the player chosen the ball that could be made to roll the fastest? Why? Why is (variable) important? Which ball could be made to roll the fastest? Why? Why is (variable) important? What about the balls that couldn't be made to go so fast? Would they all go equally slowly across the ice, or could some be made to go faster than others? Why? Why is (variable) important?

Leaning Tower scene (Outline of Galileo's famous experiment, without the results.) Do you remember the balls that you've seen in the other pictures? Suppose Galileo had dropped these balls from the Leaning Tower. Would some of them have fallen faster through the sky than others, or would they have all fallen with the same speed? Why? Which ball would have fallen the fastest? Why? Why is (variable) important? What about the other balls? Would they all fall equally slowly, or would some fall faster than others? Why? Why is (variable) important?

PROCEDURE

The materials were presented to the pupils in a quiet room at their schools. The pupils were taken to the room one-by-one and told that they would be asked questions about familiar events. It was pointed out that although comparable events might have been discussed during schoolwork, the questions were not tests of how well lessons had been mastered. The interest was simply in what pupils thought regardless of teaching. Once each pupil had consented to participate, details of name, age, gender and school class were noted and the interview commenced with presentation of the first scene and the associated questions. Roughly half the pupils in each age and gender group progressed through the four sets of scenes in the order outlined above, an order that had originally been chosen at random. The other half started with the propelled motion scenes, then had the matter transformation, moved thirdly to the object flotation, and finished with the heat transfer. The order of scenes within the topic areas remained as outlined above. It must be stressed that questioning relating to the scenes proceeded in a fashion that was only *partially* structured. Thus, although the interviewer aspired to cover the issues identified above, she did not keep the wording constant from pupil to pupil. On the contrary, she adapted the questions to suit each pupil's apparent needs and/or personal experiences. This flexibility was facilitated by the fact that, living in the area of the study and having children attending one of the schools, the interviewer was well acquainted with at least half of the sample. The pupils' responses were recorded in note form, but a tape recorder was running throughout as back-up.

CODING

It was necessary to replay the tape recordings to obtain a complete record in about 10 per cent of cases. In the remainder, it was possible to move directly from the noted responses to coding. Although coding was approached with working hypotheses (based on earlier research) as to which responses would occur, it soon became apparent that the hypotheses needed extending. Hence the final coding scheme was a mixture of a priori and post hoc decisions. The scheme was applied to the responses from thirty-one pupils by two independent coders, the interviewer and a second judge who had not been involved in the interviewing or in the design of the study. Interjudge agreement averaged 83.06 per cent, although there was variability within and between scenes. After disputed codings had been resolved by discussion, the second judge proceeded to code the remaining schedules. By utilising a judge who had not conducted the interviews, it was possible to ensure that all coding was done in ignorance of the pupils' gender and age.

The coding scheme is presented in its entirety below, together with full details of the interjudge agreement.

Heat transfer outcomes   (Interjudge agreement = 96.77%.) The pupils were categorised as follows: (a) Pans: Those who accepted or denied that switching on the cooker would cause the water to heat; (b) Forks: Those who accepted or denied that exposure to fire could cause fingers to burn; (c) Saucers: Those who accepted or denied that some water would evaporate/go into the air; d) Wraps: Those who accepted or denied that exposure to sunshine would cause ice cream to melt.

Heat transfer variables   (Interjudge agreement = 81.45%.) (a) The pupils were categorised into those who accepted or denied that the type of pan, the type of fork, the type of saucer, or the type of wrap makes a difference; (b) The different variables that each pupil used for each of the scenes were listed, with the variables flagged for scientific relevance or irrelevance e.g. *That pan's best because it's strongest* (Variable = Strength helps heating/Irrelevant); *A longer fork would take longer to burn* (Variable = Length inhibits burning/Relevant). It should be noted that irrelevant variables included those that referred to irrelevant dimensions (e.g. *Newness helps burning*) and those that used relevant dimensions in the wrong direction (e.g. *Thickness helps heating*).

Heat transfer mechanisms   (Interjudge agreement = 72.58%.) The pupils were categorised as follows: (a) Level 0: Those who did not recognise heat transfer. e.g. *The heat's inside the water*; (b) Level 1: Those who recognised heat transfer, but had no understanding of the transmissive nature of transfer e.g. *The heat's coming up from where the fire is*; (c) Level 2: Those who recognised the transmissive nature of transfer e.g. *The heat will go up the metal and burn*.

Object flotation outcomes   (Interjudge agreement = 98.39%.) Responses to the airbed and balloons scenes were categorised as follows: (a) Level 0: No changes were identified as a function of enhanced depth or height; (b) Level 1: Person-oriented changes were identified e.g. *It'll be spookier*; (c) Level 2: Physical changes apart from pressure and density were identified e.g. *It'll get* *darker and colder*; (d) Level 3: Density and/or pressure were referred to but not properly grasped e.g. *The pressure and density will get less; It'll get heavier*; (e) Level 4: Changes in pressure and density were understood e.g. *The air pressure and density will get less as you go up*.

Object flotation variables   (Interjudge agreement = 70.16%.) (a) The pupils were categorised into those who accepted or denied that the water would make a difference to the airbed or the plants, and those who accepted or denied that the sky would make a difference to the balloons or the magic carpet; (b) The different variables that each pupil used for each of the scenes were listed, with the variables flagged for scientific relevance or irrelevance e.g. *Shallow water will help it float because the deeper you go the more mass and it gets heavier* (Variable = Shallowness helps floating/Irrelevant); *Rain will push the balloon down and wind will push it up* (Variable = Rain hinders floating and wind helps/Irrelevant).

Object flotation mechanisms   (Interjudge agreement = 85.48%.) The pupils were categorised as follows: (a) Level 0: Those who did not recognise that flotation depends on the object-fluid interplay; (b) Level 1: Those who believed that fluids or their components exert a downwards force e.g. *Deep water makes it sink because there's more to draw the weight*; (c) Level 2: Those who believed that fluids act as penetrable or impenetrable barriers e.g. *A balloon can't push through the ozone layer*; (d) Level 3: Those who believed that fluids or their components exert an upwards force e.g. *Dirt is like smoke drifting up, and so it can push things up. It's the force of the water going down that pushes her up; Salt changes the volume/weight ratio and that's what affects the upthrust of the water*.

Matter transformation outcomes   (Interjudge agreement = 98.39%). The pupils were categorised as follows: (a) Charcoal: Those who accepted or denied that when charcoal turns to ash a new substance is formed; (b) Nails: Those who accepted or denied that when nails left outside eventually rust a new substance is formed; (c) Candles: Those who accepted or denied that when wax melts a new substance is formed; (d) Jelly: Those who accepted or denied that when jelly cubes are dissolved in water a new substance is formed.

Matter transformation variables   (Interjudge agreement = 77.42%.) (a) The pupils were categorised into those who accepted or denied that changes to the charcoal scene, the kind of nail, the candles scene, or the jelly scene make a difference; (b) The different variables that each pupil used for each of the scenes were listed, with the variables flagged for scientific relevance or irrelevance e.g. *A cold room would slow melting down by bringing the heat down* (Variable = Cold environment slows melting/Irrelevant); *Putting a lid* *on the barbecue will speed up the burning by stopping the oxygen* (Variable = Lid helps burning/ Irrelevant).

Matter transformation mechanisms   (Interjudge agreement = 87.90%.) The pupils were categorised as follows: (a) Level 0: Those who did not respond or who offered a redescription e.g. *The wax gets runnier and then gets harder*; (b) Level 1: Those who believed that the substance gets broken into bits or changed in volume or weight e.g. *The rain makes the nail get broken down; As it gets burned, it disintegrates into smaller bits which are inside the charcoal*; (c) Level 2: Those who believed that the substance is transformed in some way but who offered an inadequate account of the transformation e.g. *When the jelly melts, the particles are separated and hang around in the water; Heat is breaking up the charcoal and putting sulphur dioxide in*; (d) Level 3: Those who proposed an adequate or near adequate transformation e.g. *The oxygen from the air changes the iron into iron oxide.*

Propelled motion outcomes   (Interjudge agreement = 97.85%.) The pupils were categorised as follows: (a) Level 0: Those who had no idea about how speed would change e.g. *The ball will get faster and faster as it goes across the ice*; (b) Level 1: Those who had some idea e.g. *It'll go for a while at the same speed and then it will slow down*; (c) Level 2: Those who were accurate e.g. *It'll start to slow down immediately though you can't see it.*

Propelled motion variables   (Interjudge agreement = 84.68%.) (a) The pupils were categorised into those who accepted or denied that the type of ball would make a difference with the hopscotch, curling or leaning tower, and that the type of parachute would make a difference with the parachutes; (b) The different variables that each pupil used for each of the scenes were listed, and apart from lightness or heaviness with the hopscotch and curling scenes flagged for scientific relevance or irrelevance e.g. *To slow the parachute down, you need to make it bigger* (Variable = Bigness slows parachute/Relevant); *The table tennis ball will fall the fastest because it doesn't have so much to carry* (Variable = Lightness helps speed/Irrelevant – with the leaning tower scene).

Propelled motion mechanisms   (Interjudge agreement = 77.42%.) The pupils were categorised as follows: (a) Level 0: Those who did not acknowledge any physical influence e.g. *The ball gets tired*; (b) Level 1: Those who believed that the speed of the object is determined purely by its internal properties or those who believed that drag/friction operate in the direction of motion e.g. *It'll stop when its roll has gone; The pingpong ball's air will slow it down*; (c) Level 2: Those who believed that the speed of the object is influenced by variable external forces e.g. *The bumps and cracks in the stones will slow it down; The poly bag's holding the air, but there's less air to hold as the parachute comes down*; (d) Level 3: Those who believed that the speed of the object is influenced by constant external force/s e.g. *In theory, all the balls would come at the same speed – in practice it depends on the air resistance – Galileo would have been better to have done his experiment on the moon; If there's the same amount of force, the golf ball will go faster because it's small, so there's less air resistance and lots of momentum because of the mass.*

Notes

1 EVERYDAY PHYSICS AND CONCEPTUAL STRUCTURE

1 Sears (1992) has surveyed the subject choices of a large sample of A-level candidates in England and Wales. He found that slightly under 17 per cent were
registered for physics, which in itself is low enough. However, A-levels are the
more advanced of two series of public examinations, and pupils can leave school
after the first series and/or opt to study for less exacting qualifications. In fact,
A-levels would typically only be taken by pupils aspiring to higher education. For
this reason, the proportion in any given age cohort studying physics is considerably less even than Sears' 17 per cent.
2 This carries the interesting methodological implication that if the concern is with
variables, highly familiar situations are best avoided.
3 A debate has been raging for about twenty-five years as to whether stronger feedback is also provided. Some would argue, for example, that the parents of
language learning children comment on grammatical well-formedness and/or
indicate alternative structures. The details of this debate are entirely irrelevant in
the present context for even on the strongest interpretation consensual grammar
is logically impossible (see Howe, 1993a, for further details).

2 RATIONALE FOR A DEVELOPMENTAL PERSPECTIVE

1 Piaget frequently used the word 'egocentrism' to describe the child's tendency to
see the world from his/her own perspective. It is ironic that Vygotsky accepts
egocentrism on the plane of thought when (in my view rightly) he is critical of
Piaget's extension of the concept to speech.
2 Leslie's methodology is an instance of the so-called 'habituation paradigm' which
dominates contemporary research with infants. This involves familiarisation and
decreased interest (as measured by, for example, heart rate and gaze pattern)
followed by novelty and renewed interest which varies by hypothesis according to
perceived degree of novelty.

3 TEMPERATURE CHANGE AND CHILDHOOD THEORISING

1 See Chapter 7 for a detailed discussion of density and rising.
2 In a study reported subsequent to mine, Slone et al. (1996) have clearly observed
something similar with a South African sample.
3 As an example of the Critical Ratio Test, consider the following table based on
the pans data in Table 3.4:

	0 Relevant variables	*1+ Relevant variables*
Aware of transfer	63(a)	58(b)
Not aware of transfer	5(c)	0(d)

$z = (a - d) \div$ square root of $(a + d)$

$= 63 \div 7.94$

$= 7.93$

4 Although ninety-six pupils were assigned to groups, four were absent on the day scheduled for their group task. A further four were absent at the time of the post-test. Thus, eighty-eight pupils completed all stages of the procedure, with twenty-two being associated with each task format. The data were analysed by two-way (with or without critical test, with or without rule generation) ANOVA and the reported F value relates to the critical test x rule generation interaction with, as indicated, one and eighty-four degrees of freedom.

5 ENCAPSULATED KNOWLEDGE OF HORIZONTAL MOTION

1 It is clear from this that Newton was talking about average velocity, and not say, instantaneous. However, the distinction is immaterial for present purposes, so to save space the chapter will refer to 'velocity' and 'speed' rather than 'average velocity' and 'average speed'.
2 Within Scotland, 'Highers' constitute the more advanced of two series of public examinations. They are taken when pupils are aged 16 to 17, and like A- levels in England and Wales, are the main criteria for university admission.
3 It will have been noticed by now that the terms 'mass' and 'weight' are being shunned for the more ambiguous 'heaviness'. This is deliberate. In everyday discourse, we use 'mass' in a non-Newtonian fashion, and we use 'weight' in contexts where scientists would use 'mass'. This creates impossible difficulties for a chapter which, of necessity, shifts constantly from an everyday to a received perspective. The difficulties are far worse even than those posed by 'velocity' and 'speed'. Recognising this, 'heaviness' is being used where at all possible to fudge the issue.

6 HORIZONTAL AND VERTICAL MOTION COMPARED

1 Differences within children were analysed separately from differences between, with of course a repeated-measures design.
2 Bliss *et al.* (1989) tell us that their sample spanned all age bands from the first year of an English secondary school to the sixth form. The ages are inferred from the normal age range within each band.

7 FLOTATION IN LIQUIDS AND STAGE-LIKE PROGRESSION

1 Stepans *et al.* (1986) present their results on an object-by-object basis, and this reveals variability in level as a function of object. However, this does not alter the

picture as regards age trends, so for simplicity I have averaged across objects and am reporting these results.
2 I am indebted to Andy Tolmie for proposing a cluster analysis here and for carrying it out.
3 It will be remembered that ninety-five children referred to causal mechanaisms involving the object–liquid interplay. Because some of these children referred to more than one such mechanism, the total 'no. of children' across the four relevant columns of Table 7.6 is in excess of ninety-five.
4 I am loath to place too much emphasis on unpublished pilot data. However, my colleague Jenny Low and I have recently used the Howe *et al.* (1995a) and Tolmie *et al.* (1993) procedures with motion down an incline as the topic area. The apparatus was as described in Howe *et al.* (1992b). We videotaped three groups of 9- to 10-year-olds and three groups of 11- to 12-year-olds working through tasks with a rule generation plus critical test structure. The important finding is that in this propelled motion context mechanisms were never mentioned once by any of the groups.

8 FLOTATION IN GASES OR FAILURE TO FALL

1 Also included in response type (d) were answers that equated pressure with (approximations of) density, for example *I think it gets to a certain height until the pressure of the air (or the helium inside) or the weight of the helium is the same as the air outside.*
2 If breadth had been important, the simplification detailed here would be open to the same criticisms as the child-based approaches to variable selection (as outlined in the previous chapter). The simplification is defensible because breadth is not an issue.

9 AN ACTION-BASED THEORY OF CONCEPTUAL GROWTH

1 Vygotsky seems to have regarded size of zone as a potential unit of psychometric assessment, related but superior to mental age.
2 Research by Bloom *et al.* (1980) suggests that initially children exclude causal connectives like 'because' and 'so that'. They rely on the context to convey causal meanings. Causal connectives are introduced gradually during the third and fourth years.
3 DiSessa uses the term 's-p-prims' at this point, meaning 'structural p-prims'. Since the distinction between s-p-prims and p-prims is immaterial in the present context, I have simplified to p-prims within the quotation.

10 ACTION-BASED KNOWLEDGE IN A WIDER CONTEXT

1 This point parallels a claim made in relation to Vygotsky during part I, i.e. the claim that if Vygotsky is correct linguistic experiences will protect children from having to ask questions which require a possibilistic perspective.

References

Aguirre, J.M. (1988) 'Student preconceptions about vector kinematics', *The Physics Teacher* 26: 212–216.

Albert, E. (1978) 'Development of the concept of heat in children', *Science Education* 62: 389–399.

Allen, V.L. (1975) 'Social support for nonconformity', in L. Berkowitz (ed.) *Advances in Experimental Social Psychology*, Vol. 8, New York: Academic Press.

Ameh, C. (1987) 'An analysis of teachers' and their students' views of the concept of gravity', *Research in Science Education* 17: 212–219.

Anderson, A., Tolmie, A., Howe, C., Mayes, J.T. and Mackenzie, M. (1992) 'Mental models of motion?', in Y. Rogers, A. Rutherford and P.A. Bibby (eds) *Models in the Mind: Theory, Perspective and Application*, London: Academic Press.

Andersson, B. (1980) 'Some aspects of children's understanding of boiling point', in W.F. Archenhold, R. Driver, A. Orton and C. Wood-Robinson (eds) *Cognitive Development Research in Science and Mathematics*, Leeds: Leeds University Press.

—— (1986) 'Pupils' explanations of some aspects of chemical reactions', *Science Education* 70: 549–563.

Appleton, K. (1984) 'Children's ideas about hot and cold', Learning in Science Project Working Paper No. 127, University of Waikato.

Baillargeon, R. (1994) 'Physical reasoning in young infants: seeking explanations for impossible events', *British Journal of Developmental Psychology* 12: 9–33.

Baines, E. and Howe, C.J. (1995) 'Task effects on the discourse topic manipulation skills of 4, 6 and 9 year olds', Paper presented at BPS Developmental Section Conference, Strathclyde University.

Bar, V. (1986) 'The development of the conception of evaporation', Unpublished manuscript, The Hebrew University of Jerusalem.

—— (1989) 'Children's views about the water cycle', *Science Education* 73: 483–500.

Becker, B.J. (1989) 'Gender and science achievement: a reanalysis of studies from two meta-analyses', *Journal of Research in Science Teaching* 26: 141–169.

Beveridge, M. (1985) 'The development of young children's understanding of the process of evaporation', *British Journal of Educational Psychology* 55: 84–90.

Biddulph, F. (1983) 'Students' views of floating and sinking', Learning in Science Project Working Paper No. 116, University of Waikato.

Bliss, J. and Ogborn, J. (1993) 'A common sense theory of motion: issues of theory and methodology examined through a pilot study', in P.J. Black and A.M. Lucas (eds) *Children's Informal Ideas in Science*, London: Routledge.

Bliss, J., Ogborn, J. and Whitelock, D. (1989) 'Secondary school pupils' common sense theories of motion', *International Journal of Science Education* II: 261–272.

Bloom, L., Lahey, M., Hood, L., Lifter, K. and Fiess, K. (1980) 'Complex sentences:

acquisition of syntactic connectives and the semantic relations they encode', *Journal of Child Language* 7: 235–261.

Borghi, L., De Ambrosis, A., Massara, C.I., Grossi, M.G. and Zoppi, D. (1988) 'Knowledge of air: a study of children aged between 6 and 8 years', *International Journal of Science Education* 10: 179–188.

Brook, A., Briggs, H., Bell, B. and Driver, R. (1984) 'Aspects of secondary school children's understanding of heat', CLIS Full Report, Leeds University.

Brook, A. and Driver, R. (1989) 'The development of pupils' understanding of physical characteristics of air across the age range 5–16 years', CLIS Project Report, Leeds University.

Broughton, J. (1978) 'Development of concepts of self, mind, reality and knowledge', *New Directions for Child Development* 1: 75–100.

Carey, S. (1985a) *Conceptual Change in Childhood*, Cambridge, Mass.: MIT Press.

—— (1985b) 'Are children fundamentally different kinds of thinkers and learners than adults?', in S.F. Chipman, J.W. Segal and R. Glaser (eds) *Thinking and Learning Skills*, Vol. 2, Hillsdale, N.J.: Lawrence Erlbaum Associates.

Champagne, A.B., Klopfer, L.E. and Anderson, J.H. (1980a) 'Factors influencing the learning of classical mechanics', *American Journal of Physics* 48: 1074–1079.

Champagne, A.B., Klopfer, L.E. and Solomon, C.A. (1980b) 'Interactions of students' knowledge with their comprehension and design of science experiments', University of Pittsburgh: LRDC Publication 1980/9.

Chapman, L.J. and Chapman, J.P. (1967) 'Genesis of popular but erroneous diagnostic observations', *Journal of Abnormal Psychology* 72: 193–204.

—— (1969) 'Illusory correlation as an obstacle to the use of valid psycho-diagnostic signs', *Journal of Abnormal Psychology* 74: 272–280.

Chomsky, N. (1965) *Aspects of the Theory of Syntax*, Cambridge, Mass.: MIT Press.

—— (1981) *Lectures on Government and Binding*, Dordrecht: Foris Publications.

Clark, E.V. (1993) *The Lexicon in Acquisition*, Cambridge: Cambridge University Press.

Clough, E.E. and Driver, R. (1985a) 'Secondary students' conceptions of the conduction of heat: bringing together scientific and personal views', *Physics Education* 20: 176–182.

—— (1985b) 'What do children understand about pressure in fluids?', *Research in Science and Technological Education* 3: 133–144.

—— (1986) 'A study of consistency in the use of students' conceptual frameworks across different task contexts', *Science Education* 70: 473–496.

Clough, E.E., Driver, R. and Wood-Robinson, C. (1987) 'How do children's scientific ideas change over time?', *School Science Review* 69: 255–267.

Connelly, E. (1993) 'Integration into primary school of children with learning difficulties: an investigation of the relations of young children's attitudes towards, attributions of, and interactive behaviour with peers experiencing learning difficulty', Unpublished doctoral dissertation, University of Strathclyde.

Cross, R.T. and Mehegan, J. (1988) 'Young children's conception of speed: possible implications for pedestrian safety', *International Journal of Science Education* 10: 253–265.

Daehler, M.W., Lonardo, R. and Bukatko, D. (1979) 'Matching and equivalence judgments in very young children', *Child Development* 50: 170–179.

Department for Education (1995) *Science in the National Curriculum*, London: HMSO.

Department of Education and Science (1989) *Science in the National Curriculum*, London: HMSO.

DiSessa, A.A. (1988) 'Knowledge in pieces', in G. Forman and P.B. Pufall (eds) *Constructivism in the Computer Age*, Hillsdale, N.J.: Lawrence Erlbaum Associates.

—— (1993) 'Towards an epistemology of physics', *Cognition and Instruction* 10: 105–225.

Driver, R. (1985) 'Beyond appearances: the conservation of matter under physical and chemical transformations', in R. Driver, E. Guesne and A. Tiberghien (eds) *Children's Ideas in Science*, Milton Keynes: Open University Press.

Driver, R. and Russell, T. (1982) 'An investigation of the ideas of heat, temperature and change of state of children aged between 8 and 14 years', Unpublished manuscript, University of Leeds.

Dunn, J. (1989) *The Beginnings of Social Understanding*, Oxford: Basil Blackwell.

Eckstein, S.G. and Shemesh, M. (1989) 'Development of children's ideas on motion: intuition vs. logical thinking', *International Journal of Science Education* 11: 327–336.

Erickson, G.L. (1979) 'Children's conceptions of heat and temperature', *Science Education* 63: 221–236.

—— (1980) 'Children's viewpoints of heat: a second look', *Science Education* 64: 323–336.

Erickson, G. and Tiberghien, A. (1985) 'Heat and temperature', in R. Driver, E. Guesne and A. Tiberghien (eds) *Children's Ideas in Science*, Milton Keynes: Open University Press.

Fodor, J.A. (1983) *The Modularity of Mind*, Cambridge, Mass.: Bradford Books.

Foster, S. (1993) *The Communicative Competence of Young Children*, London: Longman.

Gair, J. and Stancliffe, D.T. (1988) 'Talking about toys: an investigation of children's ideas about force and energy', *Research in Science and Technological Education* 6: 167–180.

Gelman, S.A. and Markman, E.M. (1986) 'Categories and induction in young children', *Cognition* 23: 183–209.

Gentner, D. and Gentner, D.R. (1983) 'Flowing waters or teeming crowds: mental models of electricity', in D. Gentner and A.L. Stevens (eds) *Mental Models*, Hillsdale, N.J.: Lawrence Erlbaum Associates.

Gilbert, J.K. and Watts, D.M. (1983) 'Concepts, misconceptions and alternative conceptions: changing perspectives in science education', *Studies in Science Education* 10: 61–98.

Gold, E.M. (1967) 'Language identification in the limit', *Information and Control* 10: 447–474.

Gouin-Décarie, T. (1965) *Intelligence and Affectivity in Early Childhood: An Experimental Study of Jean Piaget's Object Concept and Object Relations*, New York: International Universities Press.

Grice, H.P. (1975) 'Logic and conversation', in P. Cole and J.L. Morgan (eds) *Syntax and Semantics*, Vol. 3, New York: Academic Press.

Gunstone, R. and Watts, M. (1985) 'Force and motion', in R. Driver, E. Guesne and A.Tiberghien (eds) *Children's Ideas in Science*, Milton Keynes: Open University Press.

Gunstone, R.F. and White, R.T. (1980) 'A matter of gravity', *Research in Science Education* 10: 35–44.

—— (1981) 'Understanding gravity', *Science Education* 65: 291–299.

Halliday, D. and Resnick, R. (1988) *Fundamentals of Physics*, 3rd edn, New York: John Wiley and Sons.

Harré, R. and Madden, E.H. (1975) *Causal Powers: A Theory of Natural Necessity*, Oxford: Basil Blackwell.

Harris, P. (1991) 'The work of the imagination', in A. Whiten (ed.) *Natural Theories of Mind: Evolution, Development and Simulation of Everyday Mindreading*, Oxford: Basil Blackwell.

Hayes, P.J. (1979) 'The naive physics manifesto', in D. Mitchie (ed.) *Expert Systems in the Micro Electronic Age*, Edinburgh: Edinburgh University Press.

Hennessy, S., Twigger, D., Driver, R., O'Shea, T., O'Malley, C.E., Byard, M., Draper, S., Hartley, R., Mohamed, R. and Scanlon, E. (1995a) 'Design of a computer-augmented curriculum for mechanics', *International Journal of Science Education* 17: 75–92.

—— (1995b) 'A classroom intervention using a computer-augmented curriculum for mechanics', *International Journal of Science Education*, 17: 189–206.

Hobson, R.P. (1991) 'Against the theory of "theory of mind" ', *British Journal of Developmental Psychology* 9: 33–51.

Holland, J.H., Holyoak, K.J., Nisbett, R.E. and Thagard, P.R. (1987) *Induction: Processes of Inference, Learning and Discovery*, Cambridge, Mass.: MIT Press.

Howe, C.J. (1981) *Acquiring Language in a Conversational Context*, London: Academic Press.

—— (1991) 'Children's understanding in physics', Final Report to Nuffield Foundation.

—— (1993a) *Language Learning: A Special Case for Developmental Psychology?* Hove: Lawrence Erlbaum Associates.

—— (1993b) 'Editorial: Peer interaction and knowledge acquisition', *Social Development* 2: iii–vi.

—— (1995) 'Group work in physics: towards an inclusive curriculum', in P. Potts, F. Armstrong and M. Masterton (eds) *Learning, Teaching and Managing Schools*, Milton Keynes: Open University.

Howe, C.J., Rodgers, C. and Tolmie, A. (1990) 'Physics in the primary school: peer interaction and the understanding of floating and sinking', *European Journal of Psychology of Education* V: 459–475.

Howe, C.J., Tolmie, A. and Anderson, A. (1991) 'Information technology and group work in physics', *Journal of Computer Assisted Learning* 7: 133–143.

Howe, C.J., Tolmie, A., Anderson, A. and Mackenzie, M. (1992a) 'Conceptual knowledge in physics: the role of group interaction in computer-supported teaching', *Learning and Instruction* 2: 161–183.

Howe, C.J, Tolmie, A. and Rodgers, C. (1992b) 'The acquisition of conceptual knowledge in science by primary school children: group interaction and the understanding of motion down an incline', *British Journal of Developmental Psychology* 10: 113–130.

Howe, C.J., Tolmie, A., Greer, K. and Mackenzie, M. (1995a) 'Peer collaboration and conceptual growth in physics: task influences on children's understanding of heating and cooling', *Cognition and Instruction* 13: 483–503.

Howe, C.J., Tolmie, A. and Mackenzie, M. (1995b) 'Collaborative learning in physics: some implications for computer design', in C. O'Malley (ed.) *Computer-Supported Collaborative Learning*, Berlin: Springer-Verlag.

Inhelder, B. and Piaget, J. (1958) *The Growth of Logical Thinking*, New York: Basic Books.

Johnson, C.N. and Wellman, H.M. (1982) 'Children's developing conceptions of the mind and brain', *Child Development* 53: 222- 234.

Jones, A.T. (1983) 'Investigation of students' understanding of speed, velocity and acceleration', *Research in Science Education* 13: 95–104.

Kaiser, M.K., Jonides, J. and Alexander, J. (1986) 'Intuitive reasoning about abstract and familiar physics problems', *Memory and Cognition* 14: 308–312.

Karmiloff-Smith, A. (1992) *Beyond Modularity: A Developmental Perspective on Cognitive Science*, Cambridge, Mass.: MIT Press.

Keil, F.C. (1986) 'On the structure-dependent nature of stages of cognitive development', in I. Levin (ed.) *Stage and Structure: Reopening the Debate*, New Jersey: Ablex.

—— (1990) 'Constraints on constraints: surveying the epigenetic landscape', *Cognitive Science* 14: 135–168.

—— (1992) 'The origins of an autonomous biology', in M.R. Gunnar and M. Maratsos (eds) *Modularity and Constraints in Language and Cognition: The Minnesota Symposia on Child Psychology*, Vol. 25, Hillsdale, N.J.: Lawrence Erlbaum Associates.

Kohn, A.S. (1993) 'Preschoolers' reasoning about density: will it float?', *Child Development* 64: 1,637–1,650.

Kozulin, A. (1990) *Vygotsky's Psychology: A Biography of Ideas*, New York: Harvester Wheatsheaf.

Kruger, A.C. (1992) 'The effect of peer and adult–child transactive discussions on moral reasoning', *Merrill-Palmer Quarterly* 38: 191–211.

Kruger, A.C. and Tomasello, M. (1986) 'Transactive discussions with peers and adults', *Developmental Psychology* 22: 681–685.

Kuhn, D., Garcia-Mila, M., Zohar, A. and Andersen, C. (1995) *Strategies of Knowledge Acquisition*, Chicago: University of Chicago Press.

Kuhn, T.S. (1962) *The Structure of Scientific Revolutions*, Chicago: University of Chicago Press.

Lachter, J. and Bever, T.G. (1988) 'The relation between linguistic structure and associative theories of language learning – A constructive critique of some connectionist learning models', *Cognition* 28: 195–247.

Langford, J.M. and Zollman, D. (1982) 'Conceptions of dynamics held by elementary and high school students', Paper presented at American Association of Physics Teachers, San Francisco.

Laurendeau, M. and Pinard, A. (1962) *Causal Thinking in the Child*, New York: International Universities Press.

Leslie, A. (1984) 'Spatiotemporal continuity and the perception of causality in infants', *Perception* 13: 287–305.

McClelland, J.L., Rumelhart, D.E. and Hinton, G.E. (1986) 'The appeal of parallel distributed processing', in D.E. Rumelhart, J.L. McClelland and the PDP Research Group (eds), *Parallel Distributed Processing: Explorations in the Microstructure of Cognition*, Vol. 1, Cambridge, Mass.: Bradford Books.

McCloskey, M. (1983a) 'Naive theories of motion', in D. Gentner and A.L. Stevens (eds) *Mental Models*, Hillsdale, N.J.: Lawrence Erlbaum Associates.

—— (1983b) 'Intuitive physics', *Scientific American* 248: 113–120.

McDermott, L.C. (1984) 'Research on conceptual understanding in mechanics', *Physics Today* 37: 24–32.

Maloney, D.P. (1988) 'Novice rules for projectile motion', *Science Education* 72: 501–513.

Maratsos, M. (1992) 'Constraints, modules and domain specificity: an introduction', in M.R. Gunnar and M. Maratsos (eds) *Modularity and Constraints in Language and Cognition: The Minnesota Symposia on Child Psychology*, Vol. 25, Hillsdale, N.J.: Lawrence Erlbaum Associates.

Marioni, C. (1989) 'Aspects of students' understanding in classroom settings (age 10–17): case study on motion and inertia', *Physics Education* 24: 273–277.

Medin, D.L., Wattenmacher, W.D. and Hampson, S. (1987) 'Family resemblance, concept cohesiveness, and category construction', *Cognitive Psychology* 91: 242–279.

Mervis, C.B. and Rosch, E. (1981) 'Categorisation of natural objects', *Annual Review of Psychology* 32: 89–115.

Michotte, A. (1963) *The Perception of Causality*, London: Methuen.

Minstrell, J. (1982) 'Explaining the "at rest" condition of an object', *The Physics Teacher* 20: 10–14.

Mori, I., Kojima, M. and Tsutomu, D. (1976) 'A child's forming the concept of speed', *Science Education* 60: 521–529.

Murphy, G.L. and Medin, D.L. (1985) 'The role of theories in conceptual coherence', *Psychological Review* 92: 289–316.

Nelson, K. (1973) 'Some evidence for the cognitive primacy of categorisation and its functional basis', *Merrill-Palmer Quarterly* 19: 21–39.

—— (1983) 'The derivation of concepts and categories from event representations', in E.K. Scholnick (ed.) *New Trends in Conceptual Representation: Challenges to Piaget's Theory?*, Hillsdale, N.J.: Lawrence Erlbaum Associates.

Newman, D., Griffin, P. and Cole, M. (1989) *The Construction Zone: Working for Cognitive Change in School*, Cambridge: Cambridge University Press.

Noce, G., Torosantucci, G. and Vicentini, M. (1988) 'The floating of objects on the moon: prediction from a theory of experimental facts?', *International Journal of Science Education* 10: 61–70.

Osborne, R.J. (1980) 'Force', Learning in Science Project Working Paper No.16, University of Waikato.

Osborne, R.J. and Cosgrove, M.M. (1983) 'Children's conception of the changes of state of water', *Journal of Research in Science* 20: 825–838.

Osborne, R. and Freyberg, P. (1985) *Learning in Science*, Auckland: Heinemann.

Piaget, J. (1930) *The Child's Conception of Physical Causality*, London: Routledge and Kegan Paul.

—— (1932) *The Moral Judgment of the Child*, London: Routledge and Kegan Paul.

—— (1953) *The Origins of Intelligence in the Child*, London: Routledge and Kegan Paul.

—— (1954) *The Construction of Reality in the Child*, New York: Basic Books.

—— (1970) *The Child's Conception of Movement and Speed*, London: Routledge and Kegan Paul.

—— (1974) *Understanding Causality*, New York: Norton.

Piaget, J. and Chatillon, J.F. (1975) 'Solubilité, miscibilité et flottaison', *Archives de Psychologie* 43: 27–46.

Plunkett, K. (1995) 'Connectionist approaches to language acquisition', in P. Fletcher and B. MacWhinney (eds) *The Handbook of Child Language*, Oxford: Blackwell.

Premack, D. and Woodruff, G. (1978) 'Does the chimpanzee have a theory of mind?', *Behavioural and Brain Sciences* 1: 515–526.

Rodrigues, D.M.A.P. (1980) 'Notions of physical laws in childhood', *Science Education* 64: 59–84.

Rogoff, B. (1990) *Apprenticeship in Thinking: Cognitive Development in Social Context*, New York: Oxford University Press.

Rosch, E. and Mervis, C.B. (1975) 'Family resemblances: studies in the internal structure of categories', *Cognitive Psychology* 7: 573–605.

Ruggiero, S., Cortelli, A., Dupre, F. and Vicentini-Missoni, M. (1985) 'Weight, gravity and air pressure: mental representations by Italian middle school pupils', *European Journal of Science Education* 7: 181–194.

Russell, T., Longden, K. and McGuigan, L. (1991) *Materials: Primary Space Project Research Report*, Liverpool: Liverpool University.

Russell, T. and Watt, D. (1990) *Evaporation and Condensation: Primary Space Project Research Report*, Liverpool: Liverpool University Press.

Saxena, A.B. (1988) 'Understanding the concepts of force and acceleration', *Physics Education* 5: 78–83.

Schank, R.C. and Abelson, R.P. (1977) *Scripts, Plans, Goals and Understanding*, Hillsdale, N.J.: Lawrence Erlbaum Associates.

Scottish Office Education Department (1993) *Environmental Studies 5–14*, London: HMSO.

Scriven, M. (1962) 'Explanations, predictions and laws', in H. Feigl and G. Maxwell (eds) *Minnesota Studies in the Philosophy of Science*, Minneapolis: University of Minnesota Press.

Sears, J. (1992) 'Uptake of science A levels: an ICI and BP sponsored project for ASE', *Education in Science* 149: 30–31.

Séré, M.G. (1982) 'A study of some frameworks used by pupils aged 11 to 13 years in the interpretation of air pressure', *European Journal of Science Education* 4: 299–309.

—— (1986) 'Children's conceptions of the gaseous state prior to teaching', *European Journal of Science Education* 8: 413–425.

Shayer, M. and Wylam, H. (1981) 'The development of the concepts of heat and temperature in 10–13 year olds', *Journal of Research in Science Teaching* 18: 419–434.

Shipstone, D. (1985) 'Electricity in simple circuits', in R. Driver, E. Guesne and A. Tiberghien (eds) *Children's Ideas in Science*, Milton Keynes: Open University.

Shultz, T.R. (1982) 'Rules of causal attribution', *Monographs of Society for Research into Child Development* 47, 194: No. 1.

Siegler, R.S. and Richards, D.D. (1979) 'Development of time, speed and distance concepts', *Developmental Psychology* 15: 288–298.

Slavin, R.E. (1983) *Co-operative Learning*, New York: Longman.

Slone, M., Tredoux, C. and Bokhorst, F. (1996) 'Decalage effects for heating and cooling: a cross-cultural study', *Journal of Cross-Cultural Psychology* 27: 51–66.

Solomon, J. (1983) 'Messy, contradictory and obstinately persistent: a study of children's out-of-school ideas about energy', *School Science Review* 65: 225–229.

Spensley, F., O'Shea, T., Singer, R., Hennessy, S., O'Malley, C. and Scanlon, E. (1990) 'An alternate realities microworld for horizontal motion', CITE Report No. 105, Open University.

Springer, K. (1990) 'In defense of theories', *Cognition* 35: 293- 298.

Stavy, R. and Berkowitz, B. (1980) 'Cognitive conflict as a basis for teaching quantitative aspects of the concept of temperature', *Science Education* 64: 679–692.

Stead, K. and Osborne, R. (1980) 'Friction', Science Project Working Paper No. 19, University of Waikato.

—— (1981) 'What is friction? Some children's ideas', *The Australian Science Teachers' Journal* 27: 51–57.

Stepans, J.I., Beiswenger, R.E. and Dyche, S. (1986) 'Misconceptions die hard', *The Science Teacher* 53: 65–69.

Tiberghien, A. (1980) 'Studies of classroom based research into pupils' conceptual framing of scientific ideas', in W.F. Archenhold, R. Driver, A. Orton and C. Wood-Robinson (eds) *Cognitive Development Research in Science and Mathematics*, Leeds: Leeds University Press.

—— (1984) 'Critical review on the research aimed at elucidating the sense that the notions of temperature and heat have for students aged 10 to 16 years', Proceedings of the First International Workshop: Research on Physics Education.

Tolmie, A. and Howe, C.J. (1993) 'Gender and dialogue in secondary school physics', *Gender and Education* 5: 191–209.

Tolmie, A., Howe, C.J., Mackenzie, M. and Greer, K. (1993) 'Task design as an influence on dialogue and learning: primary school group work with object flotation', *Social Development* 2: 183–201.

Twigger, D., Byard, M., Driver, R., Draper, S., Hartley, R., Hennessy, S., Mohamed, R., O'Malley, C., O'Shea, T. and Scanlon, E. (1994) 'The conception of force and motion of students aged between 10 and 15 years: an interview study designed to guide instruction', *International Journal of Science Education* 16: 215–229.

Uzgiris, I.C. and Hunt, J.McV. (1975) *Assessment in Infancy: Ordinal Scales of Psychological Development*, Urbana: University of Illinois Press.

Vygotsky, L.S. (1962) *Thought and Language*, Cambridge, Mass.: MIT Press.

—— (1978) *Mind in Society: The Development of Higher Psychological Processes*, Cambridge, Mass.: Harvard University Press.

Wattenmacher, W.D., Nakamura, G.V. and Medin, D.L. (1988) 'Relationships between similarity-based and explanation-based categorisation', in D.J. Hilton (ed.) *Contemporary Science and Natural Explanation*, Brighton: The Harvester Press.

Watts, D.M. (1982) 'Gravity – don't take it for granted', *Physics Education* 17: 116–121.

—— (1983) 'Some alternative views of energy', *Physics Education* 18: 213–217.

Wellman, H.M. (1988) 'First steps in the child's theorising about the mind', in J. Astington, P. Harris and D. Olson (eds) *Developing Theories of the Mind*, New York: Cambridge University Press.

—— (1990) *The Child's Theory of Mind*, Cambridge, Mass.: Bradford Books.

Wells, G. (1985) *Language Development in the Preschool Years*, Cambridge: Cambridge University Press.

Wertsch, J.V. (1985) *Vygotsky and the Social Formation of Mind*, Cambridge, Mass.: MIT Press.

Whitaker, R.J. (1983) 'Aristotle's not dead: student understanding of trajectory motion', *American Journal of Physics* 51: 352–358.

Whiten, A. and Perner, J. (1991) 'Fundamental issues in the multidisciplinary study of mindreading', in A. Whiten (ed.) *Natural Theories of Mind: Evolution, Development and Simulation of Everyday Mindreading*, Oxford: Basil Blackwell.

Wilkening, F. (1981) 'Integrating velocity, time, and distance information: a developmental study', *Cognitive Psychology* 13: 231–247.

Winograd, T. (1982) *Language as a Cognitive Process*, Reading, Mass.: Addison-Wesley.

Wiser, M. and Kipman, D. (1988) 'The differentiation of heat and temperature', Paper presented at the Annual Meeting of the American Educational Research Association.

Wood, D., Bruner, J. and Ross, G. (1976) 'The role of tutoring in problem-solving', *Journal of Child Psychology and Psychiatry* 17: 89–100.

Yates, J. (1990) 'What is a theory? A response to Springer', *Cognition* 36: 91–96.

Yates, J., Bessman, M., Dunne, M., Jertson, D., Sly, K. and Wendleboe, B. (1988) 'Are conceptions of motion based on a naive theory or on prototypes?', *Cognition* 29: 251–275.

Index